S0-CAH-208

Methods of
QUANTUM CHEMISTRY

Leningrad. Universitet.

Methods of QUANTUM CHEMISTRY

M. G. VESELOV
LENINGRAD STATE UNIVERSITY
LENINGRAD, U.S.S.R.

Translated by SCRIPTA TECHNICA, Inc.

TRANSLATION EDITOR
S. CHOMET
KING'S COLLEGE
LONDON, ENGLAND

1965

ACADEMIC PRESS New York • London

ACADEMIC PRESS INC.
111 Fifth Avenue, New York, New York 10003

United Kingdom Edition published by
ACADEMIC PRESS INC. (LONDON) LTD.
Berkeley Square House, London W.1

LIBRARY OF CONGRESS CATALOG CARD NUMBER: *65-28622*

Originally published as *Voprosy Kvantovoy Khimii*
by the Leningrad State University Press, Leningrad, 1963

PRINTED IN THE UNITED STATES OF AMERICA

FOREWORD

The present volume consists of five sections dealing, in varying detail, with some of the methods and trends of quantum chemistry, i.e. with the theory of the electronic structure of molecules.

The first section (by T. K. Rebane) is based on a paper which he read to the First Soviet Conference on Quantum Chemistry (Leningrad, 1961). This section gives a critical review of the computational methods for molecular properties, based on the comparison of the two limiting cases, i.e., the "united atom" and the "atoms in molecules" approaches. The former concept makes use of wave functions and energies of an atom whose nuclear charge is equal to the sum of the charges of atoms making up the molecule. The "atoms in molecules" method (proposed by Moffit, 1951), on the other hand, is based on the use of wave functions and energies of isolated atoms. Rebane also discusses the semi-empirical variants of both methods, and indicates the sources of possible errors in their application.

The following section (by M. M. Mestechkin) is devoted to the LCAO MO method, in which one-electron molecular wave functions (molecular orbitals) are constructed by linear combination of one-electron atomic wave functions (atomic orbitals). Of all quantum-chemical methods dealing with the theory of complex molecules the LCAO treatment has received the most attention. Its various forms are still in the process of development.

This method is discussed both in its most rigorous form derived from the Hartree-Fock self-consistent field method (Roothaan, 1951), and in its semi-empirical modifications. A considerable part of this exposition is based on the author's own work.

One of the most important problems of quantum chemistry is the description of elementary chemical reactions. This, however, still remains to be accomplished. Theoretical studies on reactivity of molecules are largely aimed at the development of the so-called reactivity indices, i.e., quantitative criteria giving a relative measure of the activation energies at different locations in the molecule during a reaction. The LCAO method has been widely applied to the calculations of the reactivity indices.

In the third section, Tupitsyn and Adamov survey in some detail the application of the MO method to the theory of reactivity, giving numerous examples of various types of reactions. This survey will be of interest to both theoretical and experimental chemists.

The development of the quantum theory of many-body systems has in the last few years been characterized by the successful application of new mathematical techniques. The development of such techniques as second quantization, the diagram method, and the construction and use of Green's functions, which were originally worked out in quantum field theory, has led to considerable successes in solid-state and plasma theories. In Section IV, based on a lecture he gave at the First Soviet Conference on Quantum Chemistry, E. E. Nikitin uses second quantization and Feynman's diagrams to describe the energy spectra of long molecules.

In the last section, Rebane shows how the finite-difference method can be effectively used to calculate molecular energy levels and molecular orbitals of electrons in molecules with regular structure. The following cases are examined: chain and cyclic molecules composed of identical atoms, with full equalization of bonds; the same molecules with atomic Coulomb integrals and bond resonance integrals with alternating signs; cyclic molecules, with alternation in the signs of the above integrals, in magnetic fields; polyacenes; finite atomic chains and the appearance of local states in such chains owing to disturbances of periodicity; and energy bands in some two-dimensional structures.

Although it was naturally impossible to cover all the relevant topics, it is hoped that the present volume will be a contribution to the literature on quantum chemistry.

CONTENTS

Chapter I

SEMI-EMPIRICAL METHODS IN THE THEORY OF SMALL MOLECULES

T. K. Rebane

The total energies of many-electron atoms and molecules (relative to the fully ionized atom or the fully ionized products of dissociation of the molecule) can now be calculated to an accuracy of the order of 1%. Such an order of accuracy in total energy is, however, insufficient for the calculation of the dissociation energies or the excitation energies of electronic transitions. This is because the dissociation and excitation energies appear in theoretical estimates as small differences between large approximate values, and thus the error in their calculation may range from tens to hundreds of percents. This considerably reduces the practical value of theoretical calculations for many—electron molecules, and shows why it is desirable—and sometimes necessary—to resort to semi-empirical methods.

We shall now review the present state of certain representative semi-empirical methods used in the theory of small molecules. Some of the problems are also briefly treated in [21].

In the adiabatic approximation, the total energy of a molecule consists of the energies of electrons moving in the field of fixed nuclei, the Coulomb repulsion energy of the bare nuclei, as well as the rotational and vibrational energies of the nuclear framework of the molecule. The biggest calculation difficulty arises with the electron energy, the Coulomb repulsion between the nuclei being given by the assumed nuclear configuration of the molecule, while the consideration of the rotational and vibrational energies can be deferred to a subsequent stage of calculation.

The electronic energy of a molecule is a continuous function of the internuclear distance R. As $R \to 0$, the electron cloud of the molecule becomes the electron cloud of a united atom, whose charge is equal to the sum of the charges of atoms constituting the molecule. As $R \to \infty$, on the other hand, the electron cloud splits into the electron clouds of the isolated products of dissociation (separated atoms). Between these limits, the electronic energy of the molecule is equal to the energy of those states (of united or separated atoms) which correlate adiabatically with the examined electronic state of the molecule. This correlation is the basis of two semi-empirical methods of calculating the electronic energy of small molecules: these are the "united-atom" and the "atoms in molecules" methods.

1. THE "UNITED ATOM" TREATMENT

In this approach the electronic wave function Ψ of a molecule with N nuclei is expanded into a series of orthogonal wave functions Φ_r of a united atom whose nuclear charge is .

$$Z = \sum_{k=1}^{N} Z_k, \tag{1}$$

where the center of the united atom coincides with the "center of gravity" of the positive nuclear charges of the molecule. The functions Φ_r satisfy the equations

$$H^{(0)}\Phi_r = E_r^{(0)}\Phi_r, \tag{2}$$

where $H^{(0)}$ is the Hamiltonian of the united atom, and $E_r^{(0)}$ are the experimental values of the total energy of the united atom.

A matrix H for the electronic energy operator is then constructed from the functions Φ_r, where the Hamiltonians of the molecule and of the united atom are related by

$$H = H^{(0)} + e^2 \sum_{i=1}^{n} \left(\sum_{k=1}^{N} \frac{Z_k}{r_{ik}} - \frac{Z}{r_i} \right), \tag{3}$$

In this expression r_{ik} is the distance between the k-th nucleus and the i-th electron, while r_i represents the distance between the i-th electron and the nucleus of the united atom. Because of Eq. (3), the elements of the matrix H can be expressed in terms of the experimental terms for the united atom and of the potentials which are induced by the total electron density of the united atom at the nucleus of the united atom and at the individual nuclei within the molecule.

An attractive feature of this method is the simple relation between the Hamiltonians of the united atom and the molecule, and also the strict orthogonality of Φ_r for all R. The main disadvantage is the slow convergence of the expansion of Ψ in terms of Φ_r for R corresponding to states where the internuclear distances in molecules are at equilibrium. For this reason, the united atom viewpoint can yield reliable results only at small R, i.e. in the region where the adiabatic surfaces of the potential energy of the nuclei correspond to repulsion.

Bingel [1, 2, 3] has used the "united atom" method to study the potential curves of diatonic molecules and the potential surfaces of polyatomic molecules at small R. Direct inspection of the problem shows that expansions of the non-diagonal elements of the matrix H in powers of R begin with terms involving R^2, whereas expansions of diagonal elements are of the form

$$H_{rr} = E_r^{(0)} + E_r^{(2)} R^2 + E_r^{(3)} R^3 + \dots \tag{4}$$

In a state correlating with the r-th state of the united atom, the electronic energy of the molecule becomes equal to H_{rr}, with an accuracy which is proportional to terms behaving like R^3; in that state, the adiabatic potential energy of the nuclei is given by

$$V(R) = \sum_{k>l}^{N} \frac{Z_k Z_l e^2}{R_{kl}} + E_r^{(0)} + E_r^{(2)} R^2 + E_r^{(3)} R^3 + \dots \tag{5}$$

The coefficiencts $E_r^{(2)}$ and $E_r^{(3)}$ are determined by the electron density $\rho(0)$, its radial derivative $\rho'(0)$, and the electric-field intensity gradient at the nucleus of the united atom. Equation (5) therefore establishes a relationship between the shape of the electrostatic potential curve of the molecule at small R and those characteristics of the electronic structure of the united atom which determine the various hyperfine effects in its spectrum. Such effects include isotope shifts, quadrupole interactions, and so on. Thus, the potential curves at small R (determined from scattering experiments on H+He and He+He) were used to evaluate total electron densities in the ground states of the corresponding united atoms, that is, those of lithium and beryllium.

We should also note Bingel's studies on the geometry of hydrides CH_4, NH_3, H_2O, and HF, all of which reduce to the same united atom (Ne) at small R [3]. It was found that CH_4 is tetrahedral at all values of R. As $R \to 0$, H_2O becomes more linear and NH_3 more planar. The critical O-H and N-H bond lengths at which the water molecule assumes a linear and the ammonia molecule a planar structure are related to the electron density $\rho_{Ne}(0)$. at the nucleus of the united atom.

2. THE "ATOMS IN MOLECULES" METHOD

This treatment, proposed by Moffitt [5], is based on the development of the electronic wave function for the molecule from the wave functions describing various states of the isolated atoms, and on the use of the experimental energies of the isolated (separated) atoms.

The idea behind the "atoms in molecules" approach can be illustrated by calculating the difference between the $^3\Sigma_u^+$ and $^3\Sigma_u^-$ states of the oxygen molecule [4]. Calculation by the MO method gave this difference as ~10 eV, a value which is in very poor agreement with the actual value of 2 eV. On the other hand, qualitative considerations show that the $^3\Sigma_u^+$ state is predominantly covalent, while the $^3\Sigma_u^-$ state is primarily ionic. At $R \to \infty$, these states transform, respectively, into two neutral O atoms and into a pair of ions, O^+ and O^-. Calculation of the total energy difference between two oxygen atoms and the above pair of ions gives 20 eV (the experimental value is 11.6 eV), indicating that the main error in the calculation of the energy for the $^3\Sigma_u^+ \to {}^3\Sigma_u^-$ transition lies in the inaccurate estimate of the energies of the corresponding dissociation products. It also suggests immediately that the use of experimental energies of the dissociation products can appreciably improve the accuracy of calculations which are concerned with differences between molecular terms.

In the "atoms in molecules" method, one develops the molecular wave function in the form

$$\psi = \sum_s c_s \varphi_s, \tag{6}$$

where φ_s is the normalized antisymmetrized product of the exact wave functions of the isolated atoms, corresponding to energy W_s:

$$H^{(\infty)}\varphi_s = W_s\varphi_s, \tag{7}$$

and $H^{(\infty)}$ is obtained from the molecular Hamiltonian H in the limit $R\to\infty$. At $R\to\infty$, the functions φ_s are orthogonal, but at finite R the matrix

$$S_{rs} = \int \varphi_r\varphi_s d\tau \tag{8}$$

is non-diagonal.

The matrix element of the operator H, calculated with φ_r and φ_s, is represented in the form

$$H_{rs} = S_{rs}W_s + B_{rs}, \tag{9}$$

in which B_{rs} is the matrix element corresponding to the interactions of various atoms. Since H is Hermitian, the matrix H_{rs} can be written in a more symmetric form:

$$H_{rs} = \frac{1}{2}(SW + WS^+ + B + B^+)_{rs}, \tag{10}$$

where W is a diagonal matrix constructed from the experimental energies of the separated atoms. The electronic terms E of the molecule are equal to the roots of the infinite secular equation

$$\det(H - ES) = 0. \tag{11}$$

For specific calculations, the number of basic functions must be limited, and (since we do not know the exact wave functions) matrices B and S must be calculated using wave functions $\tilde{\varphi}_s$, which only approximately describe the states of the separated atoms.

Two semi-empirical expressions have been proposed for the matrix of the molecular Hamiltonian based on $\tilde{\varphi}_s$:

$$H = \frac{1}{2}\left(\tilde{S}W + W\tilde{S} + \tilde{B} + \tilde{B}^+\right), \tag{12}$$

$$H = \tilde{H} + \frac{1}{2}\left[\left(W - \tilde{W}\right)\tilde{S}^+ + \tilde{S}\left(W - \tilde{W}\right)\right]. \tag{13}$$

where the matrices calculated by means of $\tilde{\varphi}_s$ are denoted by $\sim$.

These formulae are not equivalent, and lead to somewhat different results [6]. Equation (13) is very convenient for the empirical modification of the matrix $\widetilde{H}$ of the energy operator, which has already been calculated earlier (for example, by superposition of valence structures).

The first applications of this method [7, 8] gave results which were on the whole satisfactory, but which did not indicate the extent of approximation basic to this procedure, since additional simplifying assumptions were involved (in particular, not all the electrons of the molecule were considered). However, those calculations on the simplest molecules which took account of all the electrons [6, 9, 10] indicated serious shortcomings of the first variant of the method. Thus, it was found that the method can give an energy value of the ground state of the molecule which is lower than the experimental, and that the agreement becomes worse as the wave functions are refined (by increasing the number of structures considered or by evaluating more precisely the parameters in the approximate analytical wave functions of the atoms or ions). The main reason for these shortcomings of the "atoms in molecules" method is apparently the fact that functions $\widetilde{\varphi}_s$ do not approximate the exact functions φ_s with sufficient accuracy.

Certain modifications aimed at a partial elimination of these drawbacks, have been proposed [10, 19]. In the modification of Hurley [14], the approximate wave function of the molecule is constructed from normalized antisymmetrized products $\widetilde{\varphi}_s(\xi_s)$ of the approximate wave functions for the separated atoms

$$\widetilde{\psi} = \sum_s c_s \widetilde{\varphi}_s(\xi_s). \tag{14}$$

The average electronic energy of the molecule at finite R is calculated from

$$\widetilde{E}_R = \int \widetilde{\Psi}_R H \widetilde{\Psi}_R d\tau, \tag{15}$$

where the subscript R denotes that c_s and ξ_s (the effective charges of the analytical functions) were selected to give minimum $\widetilde{E}_R$. The approximate energies of the separated atoms are

$$\widetilde{W}_s = \int \widetilde{\varphi}_s^{(\infty)} H^{(\infty)} \widetilde{\varphi}_s^{(\infty)} d\tau, \tag{16}$$

where ξ_s in the expressions $\widetilde{\varphi}_s$ were chosen to ensure minimum $\widetilde{W}_s$. The functions $\widetilde{\varphi}_s^{(\infty)}$ are assumed to be chosen so that the matrix $H_{rs}^{(\infty)}$ is diagonal.

The main point of Hurley's method is that the theoretical $\widetilde{E}_R$ is compared with the theoretical value of the energy corresponding superpositions of separated atomic states in which the populations of the individual separated atomic states are the same as at the value of R considered. Thus the theoretical $\widetilde{E}_R$ is no longer compared with the theoretical energy of those states of separated atoms which correlate adiabatically with a given state of the molecule. In Hurley's method the populations of states $\widetilde{\varphi}_s$ are determined by

$$n_s = \sum_t c_s \widetilde{S}_{st} c_t. \tag{17}$$

Thus, when a molecule undergoes "dissociation" into separate atoms to give the above superposition of states, the approximate change in electronic energy is

$$\Delta\widetilde{E} = \sum_s n_s \widetilde{W}_s - \widetilde{E}_R. \tag{18}$$

Since the populations of states $\widetilde{\varphi}_s^{(\infty)}$ and $\widetilde{\varphi}_s$ in Eq. (18) are equal, we can expect appreciable compensation of errors in the calculation.

This is because functions $\widetilde{\varphi}_s^{(\infty)}$ and $\widetilde{\varphi}_s$ contain errors of almost equal magnitude, due to the fact that correlation between electrons had been neglected (all the variables for the individual electrons have been completely separated). Therefore, Eq. (18) gives a more precise value of ΔE than that obtained under the assumption of a molecule dissociating into states which correlate adiabatically with the state considered. Equation (18), together with the experimental energies W_s of the separated atoms, gives the total electronic energy of the molecule in the form

$$E_R = \sum_s n_s W_s - \Delta E = \\ = \sum_{r,s} c_r \left[\widetilde{H}_{rs} + \tfrac{1}{2}\left(W_r - \widetilde{W}_s\right)\widetilde{S}_{rs} + \tfrac{1}{2}\widetilde{S}_{rs}\left(W_s - \widetilde{W}_r\right)\right] c_s. \tag{19}$$

Since the coefficients c are chosen to ensure minimum E_R, Eq. (19) shows that E_R is the eigenvalue of the matrix

$$H = \widetilde{H} + \tfrac{1}{2}\left[\widetilde{S}\left(W - \widetilde{W}\right) + \left(W - \widetilde{W}\right)S\right], \tag{20}$$

which has the same form as (13). However, in contrast to the original version of the "atoms in molecules" approach, the matrices $\widetilde{H}$ and $\widetilde{S}$ are calculated with the aid of wave functions whose parameters are chosen to give minimum $\widetilde{E}_R$, while $\widetilde{W}$ are calculated with functions ensuring minimum $\widetilde{W}_s$ (with the additional condition that the matrix $H_{rs}^{(\infty)}$ is diagonal).

This so-called method of correction for the intra-atomic correlation of electrons has been successfully used by Hurley in the calculation of numerous electronic states of the diatomic hydrides of elements of the second period [15, 16]. Over 80 electronic states were considered. The errors in the calculated energies of electronic transitions were approximately 1-2%, and those in the dissociation energies about 0.5-0.7 eV.

3. COMBINATION OF THE "UNITED ATOM" AND THE "ATOMS IN MOLECULES" VIEWPOINTS

Preuss [20] has proposed a combination of the two approaches discussed above. If Φ_S and φ_s are adiabatically correlated wave functions of the united atom and of the separated atoms respectively, then the wave function of the molecule is constructed in the form

$$\psi = \sum_s \gamma_S (\Phi_S + p\varphi_S)/(1 + p), \tag{21}$$

where p is a variable parameter; $p \to 0$ when $R \to 0$, and $p \to \infty$ at $R \to \infty$. Calculations carried out by this method for the H_2^+ ion (computed dissociation energy 1.95 eV, experimental 2.78 eV) and for the H_2 molecule (respectively 4.71 and 4.72 eV) give reasonable agreement with the experimental values. No empirical methods were used for the calculation of the H_2^+ ion because the wave functions employed were exact wave functions of H atoms and the He^+ ion. The calculation of H_2, on the other hand, which involved the spectroscopic energy value of the helium atom and the simplest expression for the wave function, of the form $\exp[-\lambda(r_1+r_2)]$, was not strictly variational. It is probable that the dissociation energy of H_2 calculated in this way can be appreciably lowered by including new components in the wave function. This would increase the discrepancy between theory and experiment.

CONCLUSIONS

The "united atom" and the "atoms in molecules" approaches represent significant attempts at developing computational quantum chemical methods based on the data provided by atomic spectroscopy. On the other hand, the use of these methods in their present form may give rise to a number of errors which are difficult to control. The following sources of error can be listed:

1. The expansion of the molecular wave function into exact wave functions of the united atom or of the separated atoms may converge very slowly at finite R. The limiting of the number of terms in the expansion may produce appreciable errors.

2. The use of experimental energies of the united atom or of the separated atoms in conjunction with approximate wave functions can lead to appreciable errors if the approximate functions do not approach the exact functions to a sufficient degree.

3. The calculating individual will tend to introduce arbitrary errors into the results since the form of the approximate wave functions used is not completely unique even though some matrix elements of the Hamiltonian of the molecule are fixed, being derived from experimental values).

4. The above methods do not always satisfy the variational principle for the energy of the ground state of the molecule. The calculations may introduce errors which will increase as the molecular wave function is refined (for example, by selection of parameters or consideration of additional structures).

These shortcomings of the present semi-empirical methods for the calculation of the electronic structure of small molecules show the need for a closer study of the foundations underlying the "united atom" and "atoms in molecules" approaches in order to determine the criteria of applicability of these methods and to estimate the possible errors involved.

REFERENCES

1. W. A. Bingel. Zs. Naturf., 12a, 599, 1957.
2. W. A. Bingel. J. Chem. Phys., 30, 1250, 1959.
3. W. A. Bingel. J. Chem. Phys., 30, 1254, 1959.
4. W. Moffitt. Proc. Roy. Soc., A210, 224, London, 1951.
5. W. Moffitt. Proc. Roy. Soc., A210, 245, London, 1951.
6. R. Pauncz. Acta Physica Hung., 4, 237, 1954.
7. W. Moffitt and J. Scanlan. Proc. Roy. Soc., A218, 464, London, 1953.

8. W. Moffitt and J. Scanlan. Proc. Roy. Soc., A220, 530, London, 1953.
9. A. Rahman. Physica, 20, 623, 1954.
10. A. C. Hurley. Proc. Phys. Soc., A68, 149, 1955.
11. A. C. Hurley. Proc. Phys. Soc., A69, 49, 1956.
12. A. C. Hurley. Proc. Phys. Soc., A69, 301, 1956.
13. A. C. Hurley. Proc. Phys. Soc., A69, 767, 1956.
14. A. C. Hurley. J. Chem. Phys., 28, 532, 1958.
15. A. C. Hurley. Proc. Roy. Soc., A248, 119, London, 1959.
16. A. C. Hurley. Proc. Roy. Soc., A249, 402, London, 1959.
17. T. Arai. J. Chem. Phys., 26, 435, 1957.
18. T. Arai. J. Chem. Phys., 26, 451, 1957.
19. T. Arai. J. Chem. Phys., 28, 32, 1958.
20. H. Preuss. Zs. Naturf., 12a, 599, 1957.
21. T. Arai. Rev. Mod. Phys., 32, 370, 1960.

Chapter II

THE STUDY OF THE ELECTRONIC STRUCTURES OF MOLECULES BY THE MOLECULAR ORBITAL METHOD

M. M. Mestechkin

The treatment of molecular orbitals by means of linear combination of atomic orbitals (LCAO MO method) has recently been widely used in the theoretical study of the electronic structure of molecules. The method has been very successful in elucidating structures of a variety of elements and compounds ranging from simple diatomic species to large organic molecules, the application to large molecules being particularly fruitful. The treatment was successful even in the case of molecules containing conjugated bond systems, and has recently also proved very satisfactory for saturated hydrocarbons.

That the LCAO MO method has distinguished itself in the above areas reflects the fact that it is being intensively developed in two main directions:

1. In small molecules (2-5 atoms) the calculations are now "completely theoretical." Thus, more and more exact solutions of the non-relativistic Schrödinger equation, in the form of a

determinant constructed from one-electron functions, are sought. The increase in accuracy is attained by increasing the number of the variational parameters, extension and refinement of the atomic orbitals used as a basis for calculation (better choice of orbital exponents, use of non-integer quantum numbers, etc.). The optimum result which can be achieved in this direction is the exact solution of the Hartree-Fock equations for the molecule. All integrals which one encounters in this approach are calculated as accurately as possible. Digital computers are generally used both to calculate these integrals and to find the variational parameters. Although, in the case of diatomic and triatomic systems, it isn't too hard to find the best Hartree-Fock solution by this method, the agreement with experimental data (for example, the bond energies) remains poor because of the limitations of the MO method itself. Further perfection of the computational techniques may thus extend the scope of this non-empirical approach, but even in the future it will be limited to molecules containing not more than 20 electrons.

The main objective of this approach is the development of the "best" functions used as a basis, standardization of the computational technique, determination of the limits of applicability of the LCAO MO method, and so on. The results of such calculations provide a useful starting point for the derivation of more exact solutions of the Schrödinger wave equation by other methods (for example, by the superposition of configurations). These results can also serve as a theoretical foundation for the many existing semi-empirical variants of the MO method.

2. In the case of larger molecules, it is the semi-empirical variants of the LCAO MO method which are used for elucidation of the structure. In spite of the relatively crude approximation provided by the one-determinant wave function used, the results of calculations done by this method are in satisfactory agreement

with experiment for many properties of the molecule. This agreement is due to the ease with which empirical parameters can be introduced into the calculation.

In this case, therefore, we obtain not the absolute but the relative characteristics of individual molecules within some very large class (for example, conjugated or saturated hydrocarbons, heterocyclics, and so on). Such results are undoubtedly of interest to the chemist. The good results obtained with this semi-empirical approach are due to the virtually complete constancy of the corrections to the correlation (i.e., corrections made to the wave function of the MO method to bring it into better correspondence with the wave function) within each given class of compounds.

The calculations can be appreciably simplified by the introduction of empirical parameters. The limiting case of such a treatment is the "simple" LCAO method. The simplicity of the calculations opens wide possibilities for the use of simple chemical models in the study of many molecular phenomena. The introduction of chemical concepts has extended the method beyond such "traditional" quantum mechanical problems as calculation of spectra to such important problems as chemical reactivity.

The above semi-empirical MO method has been used in the study of an ever-increasing number of molecular properties. Thus the method was long ago applied to calculation of heats of combustion and band spectra (frequency, oscillator strengths), and more recently to the determination of internuclear distances, dipole moments, electric polarizabilities, magnetic susceptibilities, the study of the hyperfine structure in ESR spectra, chemical shifts in nuclear magnetic resonance, polarographic potentials, ionization potentials, infrared spectra, optical rotation, carcinogenic tendencies, and so on.

Thus, although the semi-empirical approach is not based on "fundamental" properties of the constituent "particles" of the molecule, it does, nevertheless, bring out the internal relationships existing between various properties of the "bodies" in the molecule, and in many cases allows a prediction of their future behavior. The semi-empirical LCAO method is thus a very useful theory of molecular phenomena.

In the present review we shall discuss only the various methods for calculating the wave functions within the framework of the LCAO MO approach. We shall not be concerned either with discussing the concrete cases of molecular functions, or with the application of the wave functions to the calculation of various molecular properties. We shall similarly omit the discussion of the various generalizations of the theory, the "extended" Hartree-Fock method, calculations of systems with incomplete shells, and other deviations from the traditional scheme. We shall consider the derivation of the fundamental equations of the LCAO MO treatment and the subsequent computational procedure, i.e., the "theoretical" variant to the method. In addition, we shall discuss two widely used semi-empirical approaches, namely the "simple" and the "self-consistent."

THE BASIC EQUATIONS OF THE LCAO MO METHOD

The molecular orbitals method consists of the application of the Hartree-Fock equations to molecules. Fock obtained from the variational principle a general form of one-electron wave functions. He then pointed out that by using a specific form of these functions, the same method can be used to derive equations for the radial components, orbital exponents, expansion coefficients, and so on [1]. Equations for the coefficients in the expansion of MO's into AO's were obtained by Roothaan [2] for the case of closed shells. Roothaan emphasized the non-linear nature of these expressions.

Today, the numerical solution of Roothaan's equations is a standard procedure in the application of the MO method to molecules, comparable with the numerical integration of the radial functions in the calculations of atoms by the Hartree-Fock method. About 80 "theoretical" papers, dealing with calculations for specific molecules, have been published (about three-quarters of these treat diatomic molecules). A detailed list and description of these articles (up to 1960) can be found in the review by Allen and Karo [3]. The number of papers dealing with the semi-empirical method runs into hundreds.

1. WAVE FUNCTION AND ENERGY

Let us now formulate the basic propositions of the MO method.

The total wave function for a system of N electrons is developed as a normalized antisymmetrized product (Slater's determinant) of one-electron functions ψ_i, each of which depends on the three-dimensional coordinates and the spin of a single electron:

$$\Psi = (N!)^{-\frac{1}{2}} \det \{\psi_1 \psi_2 \ldots \psi_N\}. \qquad (1.1)$$

The one-electron functions $\psi_i = \psi_i(x, y, z, \sigma)$ are generally known as spin orbitals. They are assumed to be orthogonal and normalized

$$\int \bar{\psi}_i \psi_j d\tau_1 = \delta_{ij}; \quad i, j = 1, 2, \ldots, N. \qquad (1.2)$$

The principal problem is to select ψ_is which ensure that the energy of the molecule, E, is minimum

$$E = \int \bar{\Psi} \hat{H} \Psi \, (d\tau). \qquad (1.3)$$

In Eqs. (1-3), integration is carried out over the coordinates, while summation proceeds over the spins of all the electrons. The factor $\hat{H}$ is the non-relativistic energy operator which accounts only for the Coulomb interactions of the particles:

$$\hat{H}=\sum_{\mu=1}^{N}(T_\mu+V_\mu)+\frac{1}{2}e^2\sum_{\mu\neq\nu}\frac{1}{r_{\mu\nu}}, \tag{1.4}$$

where T_μ is the kinetic energy operator of the μth electron:

$$T_\mu=-\frac{\hbar^2}{2m}\nabla_\mu^2 \tag{1.5}$$

(the sign ^ is used exclusively to denote many-electron operators); V_μ is the potential energy operator for the μth electron in the field of the nuclei, being given by:

$$V_\mu=-e^2\sum_{a=1}^{k}\frac{Z_a}{r_{\mu a}}, \tag{1.6}$$

in which k is the number of nuclei contained in the molecule. Substitution of (1.1) and (1.4) into (1.3), using the rules for calculating integrals from determinant functions [4], gives

$$E=\sum_{i=1}^{N}\langle i|T+V|i\rangle+\frac{1}{2}e^2\sum_{i,j=1}^{N}\left(\langle ij\left|\frac{1}{r}\right|ij\rangle-\langle ij\left|\frac{1}{r}\right|ji\rangle\right), \tag{1.7}$$

where the matrix elements are calculated by means of functions ψ_i.

Since the energy operator is independent of the spin variables, these variables can also be separated in the one-electron functions:

$$\psi_i=\varphi_i(x,y,z)\,\eta_i(\sigma). \tag{1.8}$$

The function of the space coordinates, $\varphi_i(x, y, z)$, is usually called a molecular orbital (MO).

We must merely ensure that the total wave function is an eigenfunction of the operator of the square of the total spin of the system, and of its projection on the z-axis:

$$\hat{S}^2\Psi = s(s+1)\Psi, \tag{1.9}$$

$$\hat{S}_z\Psi = s_z\Psi. \tag{1.10}$$

Condition (1.10) can be satisfied by assuming that η_i is an eigenfunction of the operator of the projection of the spin of one electron on the z-axis

$$S_z\eta_i = \pm\frac{1}{2}\hbar\eta_i, \quad i = 1, 2, \tag{1.11}$$

η_1 is commonly denoted by α, and η_2 by β.

In general, condition (1.9) leads to the requirement that should be constructed in the form of a linear combination of Slater determinants (1.1). However, the ground state of a molecule with an even number of electrons

$$N = 2n \tag{1.12}$$

which is the case considered here, corresponds to a wave function of the singlet with $s = 0$. This wave function may be constructed by assuming that each MO is populated with a pair of electrons having opposite spins:

$$\psi_{2i-1} = \varphi_i\alpha, \quad \psi_{2i} = \varphi_i\beta, \quad i = 1, 2, \ldots. n. \tag{1.13}$$

Whether the condition $\hat{S}^2\Psi = 0$ is really fulfilled is readily checked by means of the Fock-Dirac expression for $\hat{S}^2$ [5]:

$$\hat{S}^2 = N - \frac{N^2}{4} + \sum_{l>k=1}^{N} P^{\sigma}_{kl}, \tag{1.14}$$

in which P^{σ}_{kl} is the operator representing the interchange of the spin coordinates of the kth and lth electrons, which, when acting on the determinant together with other analgous operators, is equivalent to the interchange of the spin functions in the kth and lth columns. $N/2$ transpositions of α with β within the pair in (1.13) reverse the sign of Ψ, while $\frac{1}{2}\cdot\frac{N}{2}\left(\frac{N}{2}-1\right)$ transpositions of α with α and β with β in various pairs leave Ψ unchanged. The same number of transpositions of α with β and β with α reduce Ψ to zero. Since, however, $-\frac{N}{2}+\frac{N}{2}\left(\frac{N}{2}-1\right)=-N+\frac{N^2}{4}$, it follows that $\hat{S}^2\Psi=0$.

In the so-called unlimited Hartree-Fock method the molecular orbital is found from the variational principle neglecting limitation (1.13), and a term corresponding to the given spin is then separated out from the complete wave function [6]. We shall follow the usual procedure, however.

Using (1.13) and summing over the spins in (1.7), we can write the expression for the energy in the form

$$E=2\sum_{i=1}^{n} H_i + \sum_{i,j=1}^{n} (2J_{ij}-K_{ij}). \tag{1.15}$$

where

$$H_i=\int \bar{\varphi}_i (T+V)\varphi_i dv, \tag{1.16}$$

$$J_{ij}=e^2\int \frac{\bar{\varphi}_i\bar{\varphi}_j'\varphi_i\varphi_j'}{|\vec{r}-\vec{r}'|}\, dv dv', \tag{1.17}$$

are the Coulomb integrals, and

$$K_{ij}=e^2\int \frac{\bar{\varphi}_i\bar{\varphi}_j'\varphi_j\varphi_i'}{|\vec{r}-\vec{r}'|}\, dv dv' - \tag{1.18}$$

is the exchange integral. The primes refer to the spatial arguments of the functions.

It is very convenient to write Eq. (1.15) by combining the MO's in a row:

$$\varphi\left(\vec{r}\right)=\left(\varphi_1\left(\vec{r}\right),\ \varphi_2\left(\vec{r}\right),\ \ldots,\ \varphi_n\left(\vec{r}\right)\right), \tag{1.19}$$

$$E = Tr\left(2\int \varphi^{+}\cdot(T+V)\varphi dv + + e^{2}\int \frac{2\varphi^{+}\cdot\varphi'\cdot\varphi'^{+}\cdot\varphi - \varphi^{+}\cdot\varphi\cdot\varphi'^{+}\cdot\varphi'}{|\vec{r}-\vec{r}'|}dvdv'\right) \tag{1.20}$$

or

$$E = 2\int (T+V)\varphi\cdot\varphi^{+}dv + + e^{2}\int \frac{2\varphi'\varphi'^{+}\varphi\varphi^{+} - \varphi\varphi'^{+}\varphi'\varphi^{+}}{|\vec{r}-\vec{r}'|}dvdv'. \tag{1.20'}$$

Here φ^{+} is a column which is a Hermitian conjugate of φ. The primes have the same meaning as before, and the dot (which will be omitted from now on) indicates matrix multiplication. In obtaining the second expression, we made use of the possibility of cyclic transformation under the Tr sign and omitted this sign because $\varphi\varphi^{+}$ is a number. This form obviates writing out the sums and indices, and when necessary, it is easy to revert to the usual form.

Let us introduce Fock's operator f which acts on the space arguments of the functions:

$$f(\vec{r}) = h(\vec{r}) + e^{2}\int \frac{2\varphi'\varphi'^{+}}{|\vec{r}-\vec{r}'|}dv' - e^{2}\varphi(\vec{r})\int \frac{\varphi'^{+}\cdots}{|\vec{r}-\vec{r}'|}dv', \quad h = T+V. \tag{1.21}$$

The dots indicate that the last exchange term acts on a function by changing its argument $\vec{r}$ to $\vec{r}'$ and placing it under the integral sign. The energy can now be written in the form

$$E = Tr\int \varphi^{+}\cdot(f+h)\varphi dv = \int (f+h)\varphi\cdot\varphi^{+}dv. \tag{1.22}$$

To complete the set of equations for the energy, we shall also express it in terms of a one-electron density matrix

$$\rho(\vec{r},\vec{r}') = \varphi\varphi'^{+} = \sum_{=1}^{n}\varphi_{i}(\vec{r})\bar{\varphi}_{i}(\vec{r}'), \tag{1.25}$$

$$E = 2\int h\rho_{\vec{r}'\to\vec{r}}(\vec{r}, \vec{r}')\,dv +$$
$$+ e^2 \int \frac{2\rho(\vec{r}', \vec{r}')\rho(\vec{r}, \vec{r}) - \rho(\vec{r}, \vec{r}')\rho(\vec{r}', r)}{|\vec{r} - \vec{r}'|}\,dv dv', \tag{1.24}$$

$$f(\vec{r}) = h(\vec{r}) + e^2 \int \frac{2\rho(\vec{r}', \vec{r}') - \rho(\vec{r}, \vec{r}')\ldots}{|\vec{r} - \vec{r}'|}\,dv'. \tag{1.25}$$

2. THE HARTREE-FOCK EQUATIONS

We shall now turn to the variation of energy, imposing the additional condition of orthonormality on the molecular orbitals (1.2). This condition can be written as a matrix equation:

$$\int \varphi^+ \varphi\,dv = I \tag{1.26}$$

where I is an n-dimensional unit matrix. Since the Fock operator itself depends on the MO, the variation results in an increment

$$\delta f = e^2 \int \frac{2\varphi'\delta\varphi'^+ + 2\delta\varphi'\varphi'^+}{|\vec{r} - \vec{r}'|}\,dv' - e^2\delta\varphi \int \frac{\varphi'^+ \ldots}{|\vec{r} - \vec{r}'|}\,dv' -$$
$$- e^2\varphi \int \frac{\delta\varphi'^+ \ldots}{|\vec{r} - \vec{r}'|}\,dv'. \tag{1.27}$$

If we carry out the change of variables $\vec{r} \to \vec{r}'$ and perform a cyclic transformation under the Tr sign, it is not difficult to see that

$$Tr \int \varphi^+ \delta f \varphi\,dv = Tr \int \delta\varphi^+ (f - h)\varphi\,dv + Tr \int (f - h)\varphi^+ \delta\varphi\,dv \tag{1.28}$$

In accordance with (1.22)

$$\delta E = 2Tr\left(\int \delta\varphi^+ f \varphi\,dv + \int f \varphi^+ \delta\varphi\,dv\right). \tag{1.29}$$

We now invoke the additional condition and obtain from (1.29)

$$\delta\left(\sum_{i,j=1}^{n} 2\varepsilon_{ij} \int \bar{\varphi}_j \varphi_i\,dv\right) = 2Tr\,\varepsilon\delta \int \varphi^+ \varphi\,dv =$$
$$= 2Tr\left(\int \delta\varphi^+ \varphi\,dv\,\varepsilon + \varepsilon \int \varphi^+ \delta\varphi\,dv\right), \tag{1.30}$$

where 2ε is an n-dimensional matrix of Lagrange multipliers ε_{ij}. Consequently the condition of steady-state energy leads to

$$2\,Tr\left(\int \delta\varphi^{+}\,(f\varphi-\varphi\cdot\varepsilon)\,dv+\int (f\varphi^{+}-\varepsilon\cdot\varphi^{+})\,\delta\varphi dv\right)=O. \tag{1.31}$$

Since φ and φ^{+} vary independently, we set each of the corresponding coefficients in (1.31) to zero,

$$f\varphi=\varphi\cdot\varepsilon,\quad f\varphi^{+}=\varepsilon\cdot\varphi^{+}. \tag{1.32}$$

The Hermitian conjugate of the first of these equations, $f\varphi^{+}=\varepsilon^{+}\cdot\varphi^{+}$, coincides with the second equation only if the matrix of the Lagrange multipliers is Hermitian:

$$\varepsilon=\varepsilon^{+}. \tag{1.33}$$

Equations (1.32) are the well known Hartree-Fock integro-differential equations for a closed shell. Remember that the operator f (1.21) acts on the spatial arguments of φ, and the matrix ε acts on the indices of φ. We can now write out the Hartree-Fock equations in detail:

$$\begin{gathered} h\varphi_i\left(\vec{r}\right)+\sum_{j=1}^{n} e^2\int\frac{2\bar{\varphi}_j'\varphi_j}{\left|\vec{r}-\vec{r}'\right|}\,dv'\cdot\varphi_i\left(\vec{r}\right)- \\ -\sum_{j=1}^{n} e^2\int\frac{\bar{\varphi}_j'\varphi_i'}{\left|\vec{r}-\vec{r}'\right|}\,dv'\cdot\varphi_j\left(\vec{r}\right)=\sum_{j=1}^{n}\varepsilon_{ij}\varphi_j\left(\vec{r}\right),\quad i=1,\,2,\,\ldots,\,n. \end{gathered} \tag{1.32'}$$

An important property of these equations is their invariance under a unitary transformation of the MO. Writing,

$$\varphi=\omega U,\quad UU^{+}=I, \tag{1.34}$$

(where ω is a row of new orbital functions ω_i and U is an unitary n-dimensional matrix) we notice that the one-electron density matrix (1.23) remains invariant

$$\rho\left(\vec{r},\,\vec{r}'\right)=\omega UU^{+}\omega^{+\prime}=\omega\omega^{+\prime}. \tag{1.35}$$

Thus the operator f, in accordance with (1.25), is expressed in terms of the new functions ω in the same way as it is expressed in terms of the φ. Also, in the new orbitals, Eqs. (1.32) preserve their original form which includes the transformed matrix of the Lagrange multipliers

$$f\omega=\omega\Lambda, \quad \Lambda=U\varepsilon U^{-1}. \tag{1.36}$$

Since the matrix ε can undergo unitary transformation, we can regard it as diagonal. Now, the diagonal elements ε_i are equal to the eigenvalues of the Hartree-Fock operator f and the MO are its eigenfunctions. Since eigenfunctions corresponding to different eigenvalues are orthogonal, condition (1.26) will be fulfilled automatically.

It can be shown that $E-E_i=-\varepsilon_i$, where E_i is the energy of the state which is obtained from Ψ by the removal of one electron from the ith MO. The parameters ε_i are therefore equal to the adiabatic ionization potentials with the sign reversed (Koopmans' theorem) [7].

It was first shown by Fock [1] that if f is formally regarded as constant during the variation of the MO, Eqs. (1.32) can be obtained by minimizing the sum of the ionization potentials W

$$W=Tr\varepsilon=Tr\int\varphi^+f\varphi dv,$$

since

$$\varepsilon=\int\varphi^+f\varphi dv. \tag{1.37}$$

And clearly,

$$\delta W_{(f-\text{const})}=Tr\left(\int\delta\varphi^+f\varphi dv+\int(f\varphi^+)\delta\varphi\right)=\delta E.$$

3. EQUATIONS FOR ORBITAL COEFFICIENTS AND BOND POPULATIONS

We shall now apply the main idea of the LCAO method, i.e., we shall seek the MO in the form

$$\varphi_i(\vec{r}) = \sum_{q=1}^{m} \chi_q(\vec{r})\, c_{qi}, \quad i = 1, 2, \ldots, n, \tag{1.38}$$

in which $\chi_q(\vec{r})$ are fixed, linearly independent functions forming a basis for the expansion, i.e., they are atomic orbitals (AO). Later on we will look at how they are chosen.

If we use the language of representation theory and regard the columns of the coefficients of the expansion of the MO in terms of the AO

$$c_i = \begin{pmatrix} c_{1i} \\ c_{2i} \\ \cdot \\ \cdot \\ \cdot \\ c_{mi} \end{pmatrix}, \quad i = 1, 2, \ldots, n \tag{1.39}$$

as wave functions in a new representation, then we may regard the equations of the LCAO MO method as the Hartree-Fock equations of these new wave functions. With the aid of (1.39), Eqs. (1.38) can be written compactly

$$\varphi_i = \chi c_i, \quad i = 1, 2, \ldots, n, \tag{1.40}$$

where χ is a row of AO's

$$\chi = (\chi_1 \chi_2 \ldots \chi_m). \tag{1.41}$$

Collecting columns c_i to form a matrix C of dimension $m \times n$, we may write the set of Eqs. (1.40) in the form

$$\varphi = \chi C. \tag{1.42}$$

Functions χ are assumed not to be orthogonal:

$$\int \bar{\chi}_q \chi_t dv = s_{qt}, \quad q, t = 1, 2, \ldots, m, \tag{1.43}$$

or

$$\int \chi^+ \chi dv = S. \tag{1.43'}$$

The diagonal elements of the m-dimensional matrix of overlap integrals S are generally considered to be unity (normalization).

Substituting (1.42) into (1.20), we express the energy in terms of the orbital coefficients

$$E = \operatorname{Tr} C^+ \left(2 \int \chi^+ h \chi dv + 2e^2 \int \frac{\chi^+ \chi' CC^+ \chi'^+ \chi}{|\vec{r} - \vec{r}'|} dv dv' - e^2 \int \frac{\chi^+ \chi CC^+ \chi'^+ \chi'}{|\vec{r} - \vec{r}'|} dv dv' \right) C. \tag{1.44}$$

We introduce the following m-dimensional matrices:

$$H = \int \chi^+ h \chi dv, \tag{1.45}$$

$$R = CC^+ = \sum_{i=1}^{n} c_i c_i^+, \tag{1.46}$$

$$F = \int \chi^+ f_c \chi dv = H + e^2 \int \frac{2\chi^+ \chi' R \chi'^+ \chi - \chi^+ \chi R \chi'^+ \chi'}{|\vec{r} - \vec{r}'|} dv dv'. \tag{1.47}$$

The operator f_c is determined from formula (1.21) using the best approximation for φ in the form (1.42).

We shall term R the population matrix and F the energy matrix. R corresponds to the one-electron density matrix

$$\rho(\vec{r}, \vec{r}') = \chi CC^+ \chi'^+ = \chi R \chi'^+. \tag{1.48}$$

Using (1.45)-(1.47), E can be written in the form

$$E = \operatorname{Tr} R (F + H). \tag{1.44'}$$

The condition of orthogonality of the MO, that is, Eq. (1.26), is now expressed as

$$C^{+}SC=I, \quad RSR=R. \tag{1.49}$$

The second equality in (1.49) is obtained from the first by multiplying it from the left by C and from the right by C^{+}. Dim $I = n$. Therefore

$$Tr\, SR=Tr\, C^{+}SC=Tr\, I=n. \tag{1.50}$$

Equations (1.50) and (1.48) show why it is customary to call R the population matrix. Thus, according to (1.48),

$$2\sum_{q(1)t(2)} R_{qt}\bar{\chi}_q(\vec{r})\,\chi_t(\vec{r}),$$

where the sum is taken over all AO's χ_q centered in atom 1 and all AO's χ_t centered in atom 2. This sum gives the additional electron density in the region of the bond joining atoms 1 and 2—an electron density caused by the overlap of the AO's. Here we make use of the fact that the χ are atomic orbitals, i.e., assume that they differ from zero only near their respective atoms. Equation (1.50) then gives the distribution of the total molecular electron charge $2en$ over the atoms and bonds. In particular

$$n_{1-2}=2\sum_{q(1),\,t(2)} R_{qt}S_{qt} \tag{1.51}$$

gives the number of electron pairs forming the bond between atoms 1 and 2. This quantity is generally known as population of bond 1-2 [8], (or, in the terminology of [9], the population of the overlap 1-2). The population of the atom is given by an analogous expression

$$n_1=\sum_{q(1)} R_{qq}, \quad (s_{qq}=1) \tag{1.51$'$}$$

and similarly

$$n_{1-2,\,i}=2\sum_{q(1),\,p(2)} c_{pi}\bar{c}_{qi}s_{qp} \tag{1.52}$$

represents the contribution of the ith MO to the population of bond 1-2. Equation (1.52) also allows us to say whether the ith MO is bonding or anti-bonding with respect to the given bond. Depending on the sign of the sum of $n_{t-u,\,i}$ over all the bonds, the MO's may be divided into bonding or anti-bonding.

Let us warn here against a misunderstanding which may arise in connection with the determination of R in (1.46). Even with the condition of orthogonality of the AO ($S = I$) it does not follow from (1.49) that R is equal to a unit matrix. To see this we note that the rectangular matrix C does not have an inverse, and $C^+C=I$ does not mean that $CC^+=I$. The limiting case $m = n$ corresponds to a unitary transformation of the MO, given by (1.34).

We shall now calculate the energy variation

$$\delta\varphi=\chi\delta C. \tag{1.53}$$

In view of (1.47) and (1.43′), expression (1.31) becomes

$$2\,Tr\,[\delta C^+(FC-SC\varepsilon)+(C^+F-\varepsilon C^+S)\delta C]=0. \tag{1.54}$$

Hence we obtain the Hartree-Fock-Roothaan equation

$$FC=SC\varepsilon \tag{1.55}$$

for the orbital coefficients matrix C.

Since, as before, the matrix ε can be regarded as diagonal, the solution of (1.55) reduces to the determination of the eigenvectors of the Hermitian matrix F

$$Fc_i=Sc_i\varepsilon_i, \quad i=1,\ 2,\ \ldots,\ n. \tag{1.56}$$

However, the elements of F themselves depend on coefficients c_{qi} [see (1.47)], and the system of Eqs. (1.56) is thus cubic and not linear. Its solution presents some difficulties.

The equation which corresponds to (1.55) and represents the population matrix in this case is

$$SRF = FRS \tag{1.57}$$

This expression follows from the condition that $S^{-1}FR = C\varepsilon C^{+}$ is Hermitian, and is already reduced to a system of second (not third) order equations with respect to the elements of R.

4. METHODS OF SOLUTION OF THE LCAO MO EQUATIONS

Two main methods are used for the solution of the algebraic system (1.55).

The first method is derived directly from calculations on atoms, and consists of applying the principle of "self-consistency." The first n MO's of the ground state in this approximation are obtained by substituting into Eq. (1.40) the n vectors c_i belonging to the n lowest eigenvalues of the matrix F. These n vectors are used to construct a population matrix R and the energy matrix F' in the next approximation. Then new eigenvectors, new MO's, etc., are sought for this next approximation according to (1.56), until the initial and calculated vectors approach each other to a degree considered sufficiently close (the degree of approach depends on the accuracy expected from the calculation).

This iteration method, proposed by Roothaan, has become the usual procedure for the determination of molecular wave functions in the one-electron approximation. This process, which in the final analysis reduces to repeated diagonalization of a symmetric matrix, has also been programmed for various electronic computers [10-12].

The second method [13, 15] takes into account the fact that the theory which uses the algebraic system (1.55) is simpler than the integro-differential equations (1.32). A "self-consistent" population matrix is thus directly constructed by iteration, and the calculation of the energy matrix and the imposing of self-consistency become one and the same process. In addition, the quantities used in this method have more direct physical significance than the orbital coefficients.

For simplicity's sake, we shall assume $S = I$ in discussing the second method. This does not affect the generality of the discussion, since we can always carry out the preliminary orthogonalization of the atomic orbitals. For example, we can orthogonalize these orbitals by the method of Löwdin [14]

$$\chi_0 = \chi S^{-\frac{1}{2}} \tag{1.58}$$

We introduce the new unknowns

$$C_0 = S^{1/2}C, \quad P = S^{1/2}RS^{1/2} \tag{1.59}$$

into the transformed energy matrix

$$F_0 = \int \chi_0^+ f_c \chi_0 \, dv = S^{-\frac{1}{2}} F S^{-\frac{1}{2}} \tag{1.60}$$

We then replace relationships (1.44), (1.46), (1.49), (1.50), (1.55), and (1.57) by the following expressions

$$\begin{gathered} F_0C_0 = C_0\varepsilon, \quad PF_0 = F_0P, \quad E = Tr\, P(H_0 + F_0), \\ C_0^+ C_0 = I, \quad P^2 = P, \\ C_0C_0^+ = P, \quad Tr\, P = n. \end{gathered} \tag{1.61}$$

These are obtained if we formally assume that $S = I$ in the above-listed equations. The factor P is known as the bond order matrix.

The second group of equations in (1.61) allows us to assert that P is the matrix of projection onto the n eigenvectors of F_0, which are associated with n least eigenvalues of F_0. P can be expressed as a funtion of F_0, for example, as

$$P = \frac{1}{2}\left[I - F_a\left(F_a^2\right)^{-\frac{1}{2}}\right], \quad F_a = F_0 - aI, \tag{1.62}$$

where the constant a assumes the role of a chemical potential and is determined from the condition of conservation of the number of particles ($Tr\,P = n$); $(F_a^2)^{-\frac{1}{2}}$ is a matrix whose eigenvalues are equal to the moduli of eigenvalues of F_a^{-1}. We can use any iteration process to extract the root, and we can then construct P directly.

Another approach uses the method of the steepest descent. In solving system (1.61) by iteration,we do not determine the vectors C_0 themselves but find values of δC_0 which are correction factors for C_0. From Eq. (1.54) it is seen that the largest reduction in the energy is obtained if δC_0 is selected in such a way that

$$Tr\,\delta C_0^+ \left(F_0 C_0 - C_0 \varepsilon\right) \tag{1.63}$$

is minimum. Since the matrix multiplication trace can be regarded as scalar product of vectors, obtained by "extending" the matrices into rows, the method of the steepest descent [16] gives a correction which is anti-parallel to the residual "vector"

$$\delta C_0 = -\lambda\left(F_0 C_0 - C_0 \varepsilon\right). \tag{1.64}$$

The positive constant λ, which determines the length of the step, is found from the second approximation correction for the energy. Taking into account the fact that according to (1.37)

$$\varepsilon = C_0^+ F_0 C_0, \tag{1.65}$$

we have

$$\delta C_0 = -\lambda\left(I - P\right) F_0 C_0 = -\lambda L C_0, \tag{1.66}$$

where

$$L = (I - P)\, F_0 P.$$

However, the more exact matrix

$$T = C_0 + \delta C_0 = (I - \lambda L) C_0 \tag{1.67}$$

obtained in this way does not satisfy the orthogonality condition (1.61). Its columns must be orthogonalized with the aid of the corrected matrix

$$C_1 = T(T^+ T)^{-\frac{1}{2}}. \tag{1.68}$$

The new bond-order matrix P_1 is equal to

$$P_1 = C_1 C_1^+ = T(T^+ T)^{-1} T^+. \tag{1.69}$$

Since

$$C_0^+ L = 0, \quad L^+ C_0 = 0, \quad LP = L, \tag{1.70}$$

we have

$$(T^+ T)^{-1} = C_0^+ (I + \lambda^2 L^+ L)^{-1} C_0. \tag{1.71}$$

so that, in view of (1.67) and (1.71),

$$P_1 = (P - \lambda L)(I + \lambda^2 L^+ L)^{-1} (P - \lambda L^+). \tag{1.72}$$

The above equation, from which the orbital coefficients matrix has finally been eliminated, is of fundamental importance. It relates the corrected bond-order matrix directly to the starting matrix P, and can thus be used as a basis for the iteration process.

An essentially similar calculation procedure has been used in the semi-empirical method [17] and in programs for electronic computers [18].

5. SELECTION OF THE FUNCTIONS WHICH SERVE AS THE BASIS

The rate of convergence of the series (1.38) to an exact solution of the Hartree-Fock equations is determined mainly by the successful selection of the functions χ which serve as the basis for calculations. The number of such functions should, in general, be infinite.

For this reason, one of the main problems in numerical calculations done in the last dozen years has been the determination of that limited number of atomic orbitals which would give the accuracy required in practice.

The second factor in the selection of the basis functions is whether they afford the possibility of calculating the matrix elements F_{pq} with sufficient accuracy:

$$F_{pq} = H_{pq} + \sum_{s,\,t=1}^{m} R_{st}\,(2\,\langle pt \mid qs\rangle - \langle pt \mid sq\rangle). \qquad (1.73)$$

Here

$$\langle pt \mid qs\rangle = e^2 \int \frac{\bar{\chi}_p \bar{\chi}'_t \chi_q \chi'_s}{\left|\vec{r} - \vec{r}'\right|}\, dv\, dv' \qquad (1.74)$$

is the electron interaction integral. When the basis is enlarged, the number of such integrals increases extremely rapidly (approximately in proportion to m^4). The matter is further complicated by the fact that we have not yet found analytical formulae for the 3- and 4-centered integrals for the AO's which are actually used. For this reason, the above-mentioned programs for the computation of wave functions [10-12] include the preliminary calculations of these two-electron integrals. The applicability of these programs is limited to diatomic [10, 11] and linear tri- and tetra-atomic molecules [12]. The integrals necessary in these cases (with functions of the Slater type) are obtained from available tables of functions [19-22].

While numerical solutions of the Hartree-Fock equations for atoms frequently served as the basis for the earlier calculations, today practically the only functions which serve as the basis are the Slater orbitals

$$S_{nlm} = R_{nl}(r)\, Y_{lm}(\theta,\ \varphi), \qquad (1.75)$$

where Y_{lm} are spherical harmonics and

$$R_{nl} = \text{const}\, r^{n-1} e^{-\zeta_{nl} \cdot r}, \qquad (1.75')$$

ζ_{nl} is an orbital exponent. The origin of the coordinates is the position of the given atomic nucleus in the molecule. Standardization of calculations using functions (1.75), as well as of terminology [23], is now being undertaken.

The minimum basis consists of those functions of type (1.75) which correspond to all sets of quantum numbers n, l, m characterizing the AO's occupied in the ground state of atoms comprising the molecule. If the basis contains several functions with the same set of quantum numbers n, l, m but differing in the orbital exponent, it is known as a basis with "multiple ζ." If one uses orbitals corresponding to a principal quantum number greater than n in the occupied states of the atoms, the basis is termed "extended." If both these generalizations are used simultaneously, the basis is both "multiple" and "extended."

Each of these cases can be further subdivided, depending on the assumed values of the orbital exponents. The majority of calculations to date have used orbital exponents equal to the ζ for free atoms. The ζ themselves can be either the "best atomic" values, i.e., values obtained for atoms by the variation method, or they can be found from Slater's semi-empirical rules (or from tables prepared by other authors). Calculations [10] show that the "best atomic" ζ do not present any advantages; thus use of values of ζ obtained by any one of the above means produces practically the same result. The use of Slater's ζ therefore appears to be the most rational.

The best results can, of course, be attained by finding ζ from the condition of minimum energy of the molecule. In practice, such minimizing is carried out by multiple repetition of the procedure (described in Section 3) for making the orbital coefficients self-consistent at several sets of exponents. The resulting points of the surface $E=E(\zeta_1, \zeta_2, \ldots)$ are then used in conjunction with the methods of numerical analysis to determine the minimum. There are also computer programs for carrying out this process in the case of diatomic molecules [10]. In the case of hydrides and homopolar molecules of the first period (of the periodic table), three to five steps in ζ are needed to reach the minimum.

The ζ found by this variational procedure may be divided into two additional types. If, in the expansion of every MO, each nodeless function (1.75) is used with the same exponent, the resulting exponents are said to be "of the type of Slater's orbitals." If, on the other hand, in the expansion of each MO the function (1.75) appears with a different exponent, then these exponents are "of the MO type." The number of variational parameters is considerably larger in the latter case.

Calculations show that no significant enlargement of the basis is necessary to obtain "good" MO's. For example, in the case of the HF_2^- ion [24], a basis which consists of fluorine orbitals with a principal quantum number 3 and two *p*-orbitals of hydrogen with optimal exponents gives an MO which is already very close to the best one-electron functions.

Although the expansion into Slater type orbitals satisfies the requirement of rapid convergence, the disadvantage of these functions is the absence of analytical formulae for three- and four-center integrals (1.74). For this reason, the method of calculation for diatomic molecules cannot be directly applied to polyatomic systems. This has served as a motivation to look for new basis functions. With the aid of a Gaussian function [25, 26]

$$\chi = x^{a_1} y^{a_2} z^{a_3} e^{-(\alpha_1 x^2 + \alpha_1 y^2 + \alpha_3 z^2)} \qquad (1.76)$$

four-center integrals can be reduced to the calculation of an auxiliary one-dimensional integral. However, expansions in Gaussians converge very slowly. Attempts have been made at eliminating these difficulties by expanding (1.75) in functions (1.76), but this approach has not, so far, been entirely successful.

All these difficulties mean that the above methods of calculation cannot, in practice, be applied to complex molecules. The further simplifications which are necessary can be introduced by using specific empirical structural characteristics of molecules.

As of now, the most successfully developed LCAO MO treatment has been that of conjugated hydrocarbons. We will now consider this approach.

THE LCAO MO TREATMENT OF π-ELECTRONS

1. THE π-ELECTRON APPROXIMATION

Many properties of conjugated hydrocarbons (planar structure, high mobility of the electrons, spectroscopic and chemical properties, and so on) are reasonably well accounted for by the so-called π-electron approximation. The fundamental assumption of this approximation is that the only motion which one needs to take into account in describing a conjugated is that of the π-electrons (i.e., electrons whose MO's are antisymmetric with respect to the z-plane of the molecule) in the field of the structure formed by σ-bonds.

We assume that each carbon participating in the conjugation is in an sp^2-hybridized state. Each carbon is linked to the neighboring atoms by σ-bonds which are in the plane of the molecule and make an angle of 120° with each other. The MO's of the σ-electrons are formed from hybridized sp^2 one-electron wave functions symmetric with respect to the z-plane, and are localized in the regions of the corresponding bonds. The fourth valence electron of the carbon, the π-electron, can shift fairly freely along the entire structure. There is therefore a certain probability of finding the π-electron near each of the atoms constituting the conjugated system. The MO's of the π-electrons are thus delocalized, and differ appreciably from zero over the whole system of conjugated bonds.

These assumptions lead to a simplification of the equations given by (1.55). Let us separate all MO's into σ- and π-orbitals.

$$\varphi = (\sigma, \pi), \tag{2.1}$$

where σ and π are rows of MO's which are respectively symmetric and antisymmetric with respect to the z-plane.

$$\varphi(x, y, -z) = (\sigma, (x, y, z), -\pi(x, y, z)). \tag{2.2}$$

The number of π-orbitals, n, is equal to $m/2$, where m is the number of atoms in the conjugated system. The number of σ-orbitals is the same as the number of bonds in the molecule.

Since the π- and σ-orbitals refer to different irreducible representations of the symmetry group of the molecule, the system (1.55) splits into equations for the orbital coefficients of π-orbitals and equations for the σ-orbitals. The internal electrons are not considered in this approximation, being regarded as fully localized at their nuclei (full screening).

In accordance with the basic postulate of the π-electron approximation, we are interested only in the π-orbitals. We shall represent these in the form of an expansion in terms of the $2p_z$ orbitals of all the atoms participating in the conjugation (these orbitals are antisymmetric with respect to the z-plane):

$$\pi = \chi C, \ \chi_q(\vec{r}) = (2p_z)(\vec{r} - \vec{r}_q). \quad q = 1, 2, \ldots, m. \tag{2.3}$$

The σ-orbitals can be expanded in terms of the symmetric s-orbitals of the carbon and hydrogen atoms and the $2p_x$ and $2p_y$ orbitals of the carbon (see, for example, [27]). However, we shall assume them to be already know, having been preselected in such a way as to allow the best possible description of the movement of the π-electrons. For example, using the above assumption of appreciable localization, we can consider that the solution of the equations for the σ-coefficients gives an MO corresponding to the bond between atoms k and l

$$\sigma_{k-l} = \frac{1}{\sqrt{2(1+s)}}(\lambda_k + \lambda_l), \tag{2.4}$$

where λ_k (if k is a carbon atom) is the hybrid sp^2-function directed along the l-k bond (if, on the other hand, k is a hydrocarbon, λ_k is the 1s function multiplied by a certain coefficient); s is an overlap integral.

We now write an equation for the orbital coefficients of π-orbitals, which has the same form as (1.55):

$$FC = SC\varepsilon. \tag{2.5}$$

Here the matrices F and S, of dimension equal to the number of atoms in the conjugated system, are calculated using the basis functions (2.3). Since

$$\varphi = (\sigma, \chi C), \quad \varphi\varphi'^{+} = \chi R \chi'^{+} + \sigma\sigma'^{+}, \tag{2.6}$$

where R is the population matrix for the π-electrons, then in agreement with (1.47) and (1.21) we have

$$\begin{aligned} F = \int \chi^{+} h \chi \, dv + 2e^2 \int \frac{\chi^{+}\sigma'\sigma'^{+}\chi}{|\vec{r} - \vec{r}'|} dv dv' - e^2 \int \frac{\chi^{+}\sigma\sigma'^{+}\chi'}{|\vec{r} - \vec{r}'|} dv dv' + \\ + e^2 \int \frac{2\chi^{+}\chi' R \chi'^{+}\chi - \chi^{+}\chi R \chi'^{+}\chi'}{|\vec{r} - \vec{r}'|} dv dv'. \end{aligned} \tag{2.7}$$

The first term gives the energy of the π-electron in the field of the nuclei (which have a charge equal to the number of valence electrons) and the second term gives the Coulomb screening by σ-electrons. The third term represents $\sigma - \pi$ exchange interaction, while the last term accounts for the Coulomb and exchange interactions between the π-electrons in the system.

It should be stressed that Eq. (2.7) is exact within the framework of the LCAO method (if we disregard the implications of the assumption of full screening by the internal electrons) and, so far, does not contain any new assumptions. Formally, matrix (2.7) has exactly the same form as (1.47) and, if we assume

$$h_{\text{eff}} = h + 2e^2 \int \frac{\sigma'\sigma'^{+}}{|\vec{r} - \vec{r}'|} dv' - e^2\sigma \cdot \int \frac{\sigma'^{+} \cdots}{|\vec{r} - \vec{r}'|} dv' \tag{2.8}$$

seems to correspond, as it were, to an examination of a completely isolated system of π-electrons.

Neglecting the exchange term for the moment, we note that in (2.4), h_{eff} can be represented as the sum of the potentials of all the atoms in the molecule.

In fact,

$$2\sigma'\sigma'^{+} = 2\sum_{k-l} |\sigma'_{k-l}|^2 = \sum_{k-l} \frac{1}{1+s} (\lambda_k'^2 + 2\lambda_k'\lambda_l' + \lambda_l'^2). \tag{2.9}$$

The sum is taken over all bonds of the molecule. If unshared electron pairs are present, we sum over them also. Considering that this expression is part of a four-center integral and using Mulliken's approximation [28], we can replace the hybrid density $\lambda_k'\lambda_l'$ by the average density $\frac{1}{2}(\lambda_k'^2 + \lambda_l'^2) \cdot s$. Then

$$2\sigma'\sigma'^{+} = \sum_{k-l} (\lambda_k'^2 + \lambda_l'^2) = \sum_{q=1}^{m} (\lambda_{q_1}'^2 + \lambda_{q_2}'^2 + \lambda_{q_3}'^2) + \sum_{a=1}^{m'} (1s)_a^2. \tag{2.10}$$

The first sum is taken over all carbon atoms and the second over all hydrogen atoms. λ_{q1}, λ_{q2} and λ_{q3} are the three hybrid sp^2-orbitals of the qth carbon. If the atom bears unshared electron pairs (as for example, nitrogen), the factor 2 appears before the corresponding AO's in the first sum. Owing to the invariance of the density during hybridization,

$$\lambda_1^2+\lambda_2^2+\lambda_3^2=(2s)^2+(2p_x)^2+(2p_y)^2. \qquad (2.11)$$

Finally, Eqs. (2.10) and (1.6) allow us to represent the effective potential as a sum of the potentials of the atomic cores.

$$h_{\text{eff}}=T+\sum_{q=1}^{m} V_q^{\text{C}}+\sum_{a=1}^{m'} U_a^{\text{H}}, \qquad (2.12)$$

where V_q^{C} is the potential of the atomic core of the carbon

$$V_q^{\text{C}}=e^2\left(-\frac{Z_q'}{|\vec{r}-\vec{r}_q|}+\int\frac{(2s')_q^2+(2p_x')_q^2+(2p_y')_q^2}{|\vec{r}-\vec{r}'|}\,dv'\right) \qquad (2.13)$$

$Z_q' = 4$ is the number of the valence electrons, and U_a^{H} is the potential of the neutral hydrogen atom

$$U_a^{\text{H}}=e^2\left(-\frac{1}{|\vec{r}-\vec{r}_a|}+\int\frac{(1s')^2}{|\vec{r}-\vec{r}'|}\,dv'\right). \qquad (2.13')$$

It is more convenient to express V_q^c in terms of the spherically symmetrical potential of the neutral carbon atom U_q^{C}, separating out the potential due to the π-electrons:

$$V_q^c=U_q^c-Z_a^{\pi}e^2\int\frac{(2p_z')_q^2}{|\vec{r}-\vec{r}'|}\,dv', \qquad (2.14)$$

$Z_q^{\pi}=1$ is the number of π-electrons.

$$U_q^c=e^2\left(-\frac{Z_q'}{|\vec{r}-\vec{r}|}+\int\frac{\alpha(r')}{|\vec{r}-\vec{r}'|}\,dv'\right)=$$

$$=-4\pi\int_r^{\infty}\left(\frac{r'}{r}-1\right)\alpha(r')\,r'dr', \qquad (2.14')$$

$$\alpha(r')=(2s')^2+(2p_x')^2+(2p_y')^2+(2p_z')^2,$$

where α depends only on the distance r' to the qth nucleus. However, for nitrogen in pyridine, $Z_k^{\pi} = 1$, but U_q^c must be increased by

$$e^2\left(-\frac{1}{r}+\int\frac{\lambda_q'^2}{|\vec{r}-\vec{r}'|}dv'\right)$$

λ_q being the hybrid sp^2 orbital of the unshared pair. For nitrogen in pyrrole $Z_q^{\pi} = 2$, and the potential increment is

$$e^2\left(-\frac{1}{r}+\int\frac{(2p_z')^2}{|\vec{r}-\vec{r}'|}dv'\right).$$

The thus separated ionic component of the potential (2.13), characterizing the attraction of π-electrons towards the core structure is almost compensated for by the repulsion from the other π-electrons. We shall therefore consider this component in conjunction with the last term of (2.7). We substitute (2.13) and (2.10) into (2.7), bearing in mind that

$$\sum_{q=1}^{m} Z_q^{\pi}\int\frac{(2p_z')_q^2}{|\vec{r}-\vec{r}'|}dv' = \int\frac{\chi' Z^{\pi}\chi'^+}{|\vec{r}-\vec{r}'|}dv',$$

where Z^{π} is a diagonal matrix composed of Z_q^{π}.

$$\begin{aligned} F = H_{\text{core}} - e^2\int\frac{\chi^+\sigma\sigma'^+\chi'}{|\vec{r}-\vec{r}'|}dv dv' + \\ + e^2\int\frac{\chi^+\chi'(2R-Z^{\pi})\chi'^+\chi - \chi^+\chi R\chi'^+\chi'}{|\vec{r}-\vec{r}'|}dv dv' = \\ = H_{\text{eff}} + e^2\int\frac{\chi^+\chi'(2R-Z^{\pi})\chi'^+\chi - \chi^+\chi R\chi'^+\chi'}{|\vec{r}-\vec{r}'|}dv dv'. \end{aligned} \tag{2.15}$$

where H_{core} is the matrix of the energy of interaction of the π-electron with the neutral structure

$$H_{\text{core}} = \int\chi^+\left(T + \sum_{k=1}^{m} U_k^{\text{C}} + \sum_{a=1}^{m'} U_a^{\text{H}}\right)\chi dv. \tag{2.16}$$

Formula (2.15) summarizes the main postulates of the π-electron approximation, and is the foundation for the π-electron calculations.

2. STRUCTURE INTEGRALS

We can now turn to a more detailed investigation of the matrix elements of F.

We shall first consider the case structure integrals of (2.16) beginning with

$$\int \bar{\chi}_q (T + U_q^{\mathrm{C}}) \chi_q dv. \tag{2.17'}$$

This equation shows the energy of an electron in the field created by a carbon represented in terms of the neutral atom plus a correction for its valence state. In other words, this expression, which characterizes the attraction of an additional electron to the carbon, should be equal to the negative electron affinity of the atom

$$\int \bar{\chi}_q (T + U_q^{\mathrm{C}}) \chi_q dv = -A_c. \tag{2.17}$$

The value $A_{\mathrm{C}} = 0.69$ electron volts [29] proposed by Mulliken [31] is assumed in the majority of existing papers [29, 30]. Analogously

$$\int \bar{\chi}_q (T + V_q^{\mathrm{C}}) \chi_q dv = -I_{\mathrm{C}}, \tag{2.18}$$

where $I_{\mathrm{C}} = 11.22$ electron volts is the ionization potential of carbon in its valence state because (2.18) represents the energy of an electron in the field of ion C^+. In view of (2.14), the following expression should hold:

$$\begin{aligned} -I_{\mathrm{C}} &= -A_{\mathrm{C}} - e^2 Z_q^{\pi} \int \frac{\chi_q^2 \chi_q'^2}{|\vec{r} - \vec{r}'|} dv dv', \\ Z_{\mathrm{C}} &= 1 \end{aligned} \tag{2.19}$$

All these considerations refer to carbon, but the formulae are written in such a way that they remain valid for other atoms participating in the conjugation. Equation (2.19) is physically compatible with (2.18) and (2.17) provided we assume that χ_p, that is, the AO of the valence electron of the carbon atom, satisfies the equation

$$(T+V_q^{\mathrm{C}})\chi_q = -I_{\mathrm{C}}. \tag{2.20}$$

The placing of an additional electron in this orbital gives the energy of an electron pair

$$-2I_{\mathrm{C}} + e^2 \int \frac{\chi_q^2 \chi_q'^2}{|\vec{r} - \vec{r}'|} dv dv',$$

i.e., according to (2.19), $-I_{\mathrm{C}} - A_{\mathrm{C}}$. This means that the addition of an extra electron lowers the total energy of the system by a factor of A_{C} in accordance with (2.17).

The expression

$$\gamma_0 = e^2 \int \frac{\chi_q^2 \chi_q'^2}{|\vec{r} - \vec{r}'|} dv dv' = I_{\mathrm{C}} - A_{\mathrm{C}} \tag{2.21}$$

for the one-center Coulomb integral, which follows from (2.19), was obtained in [29] from somewhat different considerations, and allowed an empirical estimation of that integral. Thus,

$$\gamma_0 = 10{,}53 \text{ eV} \tag{2.21'}$$

Equations (2.17) and (2.18) impose two conditions on the parameters determining the form of χ_q. This means that there should be at least two such parameters. In order to satisfy these conditions and yet select only the orbital exponents of the $2p_z$-function, Ruedenberg [32] postulated that χ_q participates in the MO expansion and in the expression for the potential (2.14) with different exponents, ζ and ζ_{C}, respectively. Direct selection of exponents on the basis of (2.17) and (2.18) has led to $\zeta = \zeta_{\mathrm{C}} = 1$, but calculation of other core structure integrals with such specifically defined χ_q does not give satisfactory results (in particular, the resonance integral becomes positive).

Another method for finding these parameters was also tried in [32]. From the condition which minimizes expression (2.17), in which ζ_C appears as a parameter, ζ was determined as a functon of ζ_C. Equating the minimum value of the resulting integral to 0.69 electron volts, Ruedenberg obtained $\zeta_C = 1.12863045$ and $\zeta = 1.6178932$. With these orbital exponents the resonance integral assumed a value almost equal to the experimental, but the one-center integral γ_0 was unfortunately rather high in comparison to that of (2.21′).

These calculations show that, to obtain the best agreement with experiment, we must limit ourselves at the present stage of development of the method to a semi-empirical scheme. We shall postulate values for certain integrals entering into most calculations [particularly in Eqs. (2.17) and 2.18)], leaving inexact the form of the functions which enter into these equations. We may, for example, assume that these values can be obtained from the parameters of σ-functions contained in the potential U_q^C. Such empirical values are used as the basis for most π-electron calculations [17, 29, 30, 33, 18, 34]. Further calculations are therefore aimed not so much at the derivation of formulae for computing the elements of the energy matrix as at an elucidation of the structure of this matrix in order to make the most rational introduction of empirical parameters.

In addition to (2.17′), the diagonal element of H_{qq} contains integrals of the type

$$\int \bar{\chi}_q U_s \chi_q dv, \quad s \neq q. \tag{2.22}$$

Since the potential U_s of the neutral atom drops off exponentially with increasing distance from it, it is quite reasonable to limit ourselves to the "tight-bonding" approximation. We then disregard all integrals which contain exponentially decreasing functions computable in the vicinity of points distant by more than one bond length from the center. In other words, in the "tight-bonding" approximation we retain all quantities which are proportional to the overlap integral for the neighboring atoms q and t.

$$s_{qt}=\int \bar{\chi}_q \chi_t dv = s. \tag{2.23}$$

With the second method of selection, for carbon s = 0.2468. We can neglect all terms of the order of s^2 and those of the order of the overlap integrals of non-neighboring atoms.

This means that in the "tight-bonding" approximation

$$\int \bar{\chi}_q \left(\sum_{q=1}^{m} U_q^{\mathrm{C}} + \sum_{a=1}^{m'} U_a^{\mathrm{H}} \right) \chi_q dv = i_q^{\mathrm{C}} \langle U^{\mathrm{C}} \rangle + i_q^{\mathrm{H}} \langle U^{\mathrm{H}} \rangle = 3 \langle U^{\mathrm{C}} \rangle - i_q^{\mathrm{H}} \delta. \tag{2.24}$$

where $\langle U^{\mathrm{C}} \rangle$ and $\langle U^{\mathrm{H}} \rangle$ are the penetration integrals for the neighboring carbon and hydrogen, respectively;

$$\delta = \langle U^{\mathrm{C}} \rangle - \langle U^{\mathrm{H}} \rangle; \tag{2.25}$$

i_q^{H} = 1, 2, 3 is the number of hydrogen atoms which are vicinal to the qth atom of the conjugated system.

The integrals comprised in H_{tq} at $t \neq q$ are non-zero only if t and q are vicinal atoms of the conjugated systems, since the integrand similar to $\chi_t \chi_q$ and χ decreases exponentially. Therefore

$$\mathrm{H}_{tq} = \int \bar{\chi}_t \left(T + \sum_{k=1}^{m} U_k^{\mathrm{C}} + \sum_{a=1}^{m'} U_a^{\mathrm{H}} \right) \chi_q = \int \bar{\chi}_t (T + U_t^{\mathrm{C}} + U_q^{\mathrm{C}}) \chi_q dv, \tag{2.26}$$

t and q being vicinal atoms. In other cases, $H_{tq}=0$.

Using the "tight-bonding" approximation, we shall now examine the effect of that part of the energy matrix (2.15) which accounts for the exchange with σ-electrons. In a typical diagonal term,

$$e^2 \int \frac{\bar{\chi}_q \left(\sum_{i=1}^{n} \sigma_i \bar{\sigma}'_i \right) \chi'_q}{|\vec{r} - \vec{r}'|} dv dv' = 3K_q \tag{2.27}$$

we must clearly retain only the three terms of the whole sum over the σ-bonds which belong to the σ-bonds of the qth atom. K_q is the exchange integral of the σ- and π-orbitals of the same atom. Indeed, $3K = K_{xz} + \frac{1}{2} K_{sz}$, in accordance with (2.11), where K_{xz} is the exchange integral of the $2p$ orbitals of one atom, and K_{sz} is the exchange integral of the $2p$ and $2s$ orbitals. At $t \neq q$, the ith σ-bond should belong to the qth and tth atoms of the conjugated system simultaneously, and therefore

$$\int \bar{\chi}_t \sum_{i=1}^{n} \frac{\sigma_i \sigma'_i \chi'_q}{|\vec{r} - \vec{r}'|} dv dv' = L_{tq} \tag{2.28}$$

differs from zero only if t and q are neighbors. L_{tq} is the hybrid integral of the $2p_z$ orbitals of vicinal atoms joined by a σ-orbital.

The form of the matrix H_{core} is therefore unchanged by the addition of exchange terms. To summarize all the results of this section, we write the sum of the first two terms of the matrix (2.15) in the form

$$H_{\text{eff}} = \alpha + \beta, \tag{2.29}$$

where α denotes a diagonal matrix with elements

$$\alpha_q = -A_C + 3\langle U^C \rangle + 3K - i_q^H \delta, \quad q = 1, 2, \ldots, m. \tag{2.30}$$

The elements of the matrix β will be non-zero only for neighboring atoms

$$\beta_{tq} = L_{tq} + \int \chi_t (T + U_t + U_q) \chi_q \, dv. \tag{2.31}$$

In the very widely used approach [17, 29], one doesn't separate the potentials of neutral atoms but uses equation (2.20). In this case H_{eff} is also represented in the form given by (2.29), but with substitution of A_{C}, according to (2.19), and with certain changes in the form of β_{tq} in (2.31).

As far as the semi-empirical methods based on (2.29)-(2.31) are concerned, the most important conclusion is that in the matrix of effective interaction with the core, the only elements different from zero are the diagonal elements and elements pertaining to vicinal atoms in the conjugated system. These elements can be replaced by empirical parameters specific to each type of atom present in the molecule, the replacement being done for each pair of neighboring atoms. The type of an atom is determined not only by its specific chemical nature, but also by its immediate surroundings, in agreement with (2.30).

Thus, in the case of hydrocarbons, for example, the simplest assumption is that $\delta = 0$, and that the parameters of all atoms α and of all bonds β are equal, by virtue of (2.30) and (2.31). Then

$$H_{\mathrm{eff}} = \alpha I + \beta M, \tag{2.32}$$

where M is a topological matrix whose elements are different from zero and equal to 1 only for vicinal atoms. The same matrix also expresses the overlap matrix in the "tight-bonding" approximation:

$$S = I + sM, \tag{2.33}$$

where s is determined from formula (2.23), and S evidently commutes with H_{eff}.

Note that when we do not assume a "tight-bonding" approximation and replace that postulate by the assumption that $[H_{\mathrm{eff}}, S] = 0$ (S already includes all overlap integrals), the insignificant corrections are negligible [35].

3. ELECTRON INTERACTION INTEGRALS

Let us now turn to the ionic part of the energy matrix (2.15). This contains the electron interaction integrals (1.74) with $2p_z$ Slater functions. The difficulties associated with calculation of these have already been mentioned. The "tight-bonding" approximation corresponds to neglect of all integrals in which the AO's in the hybrid density $\chi_t(\vec{r})\chi_q(\vec{r})$ region refer to non-neighboring atoms. This method of accounting for electron interaction was adopted by Goeppert-Mayer and Sklar [36] for the spectrum of benzene. This was the first paper in which this interaction was taken into account in the π-electron approximation. All Coulomb integrals

$$\gamma_{tq} = \gamma(R_{tq}) = e^2 \int \frac{|\chi_t(\vec{r})|^2 |\chi_q(\vec{r'})|^2}{|\vec{r} - \vec{r'}|} dv dv', \tag{2.34}$$

all exchange integrals for neighboring AO's

$$e^2 \int \frac{\bar{\chi}_t(\vec{r})\bar{\chi}_q(\vec{r'})\chi_q(\vec{r})\chi_t(\vec{r'})}{|\vec{r} - \vec{r'}|} dv dv'$$

and all hybrid integrals (R_{tq} is the internuclear distance)

$$e^2 \int \frac{\bar{\chi}_t(\vec{r})\bar{\chi}_q(\vec{r'})\chi_t(\vec{r})\chi_t(\vec{r'})}{|\vec{r} - \vec{r'}|} dv dv'$$

were retained. These two-center integrals can be calculated analytically. However, since this is not possible for most integrals which are encountered in practice, it seems appropriate to calculate them approximately in some semi-empirical method. Since all quantities are calculated to within terms of the order of s^2, the integrals should be calculated from approximate formulae which ensure the same accuracy.

In this connection, we should like to suggest that the agreement of quantum-mechanical calculations with experiment is often due to mutual cancellation of errors. For this reason, all factors of the same order must be considered simultaneously. Such a compensation takes place to some extent in the above scheme of the LCAO MO method.

Veselov [48] has shown that the non-orthogonality integrals may serve for the estimation of both the sign and the magnitude of the electron interaction integrals. Mulliken's [28] formulae, which are also based on this idea, happen to give the values of many-center integrals with accuracy to terms in s^2. The substitution

$$\chi_p\chi_q = \frac{1}{2}\left(\chi_p^2 + \chi_q^2\right) s_{pq}, \tag{2.35}$$

is made and all integrals are reduced to Coulomb integrals

$$\langle pr \mid qs \rangle = \frac{1}{4} s_{pq} \cdot s_{rs} \cdot (\gamma_{pr} + \gamma_{ps} + \gamma_{qr} + \gamma_{qs}). \tag{2.36}$$

All electron interaction integrals can thus be calculated if the function $\gamma(R)$ is known. This function can of course be computed from (2.34), but this does not lead to agreement with experimental values, especially at small R. In particular, the value for $\gamma(0)$ is about 6 electron volts higher than 10.53 electron volts. The best way of avoiding this discrepancy is to construct $\gamma(R)$ semi-empirically.

We shall not, however, write out the detailed form of the energy matrix based on these assumptions. The point is that if we turn from the population matrix R to the bond-order matrix P(1.59), and make use of the normalized energy matrix F_0, then in the "tight-bonding" approximation the F_0 matrix depends on P in the same way as F depends on R in the approximation of zero differential overlap [29, 33]. Let us now proceed to the examination of this point.

As has been pointed out, equations containing P are formally obtained by setting $S = I$. This corresponds to the assumption

$$\chi_t \chi_q = 0 \text{ for } t \neq q, \tag{2.37}$$

which, on the assumption of a zero differential overlap can also be applied to the matrix of the electron interaction integrals. We shall accept this assumption with respect to the normalized energy matrix F_0. Consequently, the expression for the latter is obtained by applying (2.37) to (2.15), i.e., replacing $\chi^+\chi$ by the diagonal matrix D

$$D = [|\chi_1|^2, |\chi_2|^2, \ldots, |\chi_m|^2]. \tag{2.38}$$

In fact, according to (1.59) and (1.60),

$$F_0 = S^{-\frac{1}{2}} H_{\text{eff}} S^{-\frac{1}{2}} + e^2 \int \frac{2X' \text{Tr} PX' - XPX' - X \,\text{Tr}\, D'Z^{\pi}}{|\vec{r} - \vec{r}'|} dv dv',$$

where the matrix

$$X = S^{-\frac{1}{2}} \chi^+ \chi S^{-\frac{1}{2}}$$

can be replaced by the matrix D in Mulliken's approximation (2.34), using

$$S^{-\frac{1}{2}} = I - \frac{1}{2} sM$$

We shall keep the expression for the first term of (2.29) as before. However, we will not attribute major significance to the formulae for α_p and β_{pq}, but merely regard these quantities as empirical parameters. In fact, if we reject the term $\beta s M^2$, which expresses the influence of second-order neighbors, we then have

$$S^{-\frac{1}{2}} H_{\text{eff}} S^{-\frac{1}{2}} = \alpha I + \beta' M,$$

where $\beta' = \beta - \alpha S$ since $\beta \sim s$. Practically the same result is obtained [30] (we no longer have full correspondence to zero differential overlap) if the coefficients a and b in $S^{-\frac{1}{2}} = aI + bM$ are selected on the basis of the conditions of the best approximation. Thus,

$$F_0 = H_{\text{eff}} + e^2 \int \frac{D\,Tr(2P - Z^{\pi})\,D' - DPD'}{|\vec{r} - \vec{r}'|}\,dvdv'. \tag{2.39}$$

Individual elements are of the form

$$F^0_{qq} = \alpha_q + (P_{qq} - Z^{\pi}_q)\,\gamma_{qq} + \sum_{\substack{t=1\\(t\neq q)}}^{m} (2P_{tt} - Z^{\pi}_t)\,\gamma_{qt}, \tag{2.40}$$

$$F^0_{qt} = \beta_{qt} - P_{qt}\gamma_{qt}, \; q \text{ and } t \tag{2.41}$$

are neighboring atoms. In all other cases,

$$F_{qt} = -P_{qt}\gamma_{qt}. \tag{2.42}$$

The expression for the diagonal element can also be written as

$$F^0_{qq} = \alpha^*_q - W_q + \left(P_{qq} - \frac{1}{2} Z^{\pi}_q\right)\gamma_{qq} + \sum_{\substack{t=1\\(t\neq q)}}^{m} (2P_{tt} - Z^{\pi}_t)\,\gamma_{qt}, \tag{2.43}$$

where

$$\alpha^*_q = 3\,(\langle U^{\mathrm{C}}\rangle + K) - i^{\mathrm{H}}_q\,\delta$$

is a very small quantity, and

$$W_q = \frac{I_q + A_q}{2} \tag{2.44}$$

is Mulliken's electronegativity.

Formulae (2.41)-(2.43) summarize all the assumptions of the π-electron approximation. Since most calculations on conjugated systems have been carried out using the energy matrix in the form shown above, their results can be used to assess the value of this for of the LCAO MO method. The satisfactory agreement with experiment which has been obtained for a whole series of properties of a great number of molecules is the best evidence for the validity of the assumptions involved in (2.41)-(2.43).

Thus, in order to carry out computations by this traditional scheme of "self-consistent" calculations of conjugated molecules, we must have at our disposal a set of parameters α_p, β_{pq}, and a function $\gamma(R)$ (the geometry of the molecule is assumed to be known).

The empirical methods of obtaining these quantities have been developed to the greatest extent in the case of hydrocarbons. Let us note here that the numerical value of α does not affect the calculation of orbital coefficients C_0 and bond orders P, since, according to (1.61), F^0 and $F^0-\alpha I$ have the same eigenvectors. We are thus left with the parameter β and the function $\gamma(R)$.

Fortunately, among the unsaturated hydrocarbons there is an example (benzene) in which the orbital coefficients (bond orders) are fully determined by the symmetry of the molecule. Benzene is therefore very suitable for testing various assumptions concerning $\gamma(R)$ and β. These two quantities are chosen so as to give the best agreement with the experimental data for benzene (spectral data [29, 37] or the heat of combustion [33]).

The simplest assumption regarding $\gamma(R)$ is that $\gamma(R)\sim\frac{1}{R}$, assuming $\gamma(0)$ to be finite [33].

In [29], $\gamma(R)$ is calculated at large distances as the energy of the Coulomb interaction of two uniformly charged bodies formed by two spheres tangent to each other. This structure replaces the dumbbell-shaped $2p_z$ orbital. The radii of these spheres are selected so as to obtain, at large R, the best agreement with the results of calculations which use Eq. (2.34) with Slater's $2p_z$ functions. The interpolation formula $\gamma=\gamma_0+bR+cR^2$ was used at small R; γ_0 was determined from (2.21), and b and c were determined by matching conditions. Slight modifications of this method can be found in [38] and [39].

The best results are not obtained by second-order interpolation at small R, but by direct determination of $\gamma(R)$ from the spectral data for benzene. The three experimentally determined intervals between excited levels in benzene are directly expressed by three differences $\gamma(R_i) - \gamma_0$, where R_i are three possible interatomic distances in the benzene molecule. The resonance integral β is omitted from these formulae. It is thus possible to find three values of $\gamma(R)$ at small R. This method of constructing $\gamma(R)$ has been adopted by Pariser [40] and Ruedenberg [37], and it is the most perfect of the semi-empirical methods. The above authors differed slightly in their approach. In [40], $\gamma(R)$ was calculated for large R with Slater functions, while in [37] it was approximated by the expression

$$\gamma(R) = \frac{10{,}27244}{R}\left(1 - e^{-\alpha R} \cdot \sum_{k=0}^{5} a_k R^k\right) \text{eV}. \tag{2.45}$$

The constants a_k in the above formula were determined from the condition that $\gamma(R)$ pass through $\gamma(0), \gamma(R_i)$; $i = 1, 2, 3$, and $\gamma'(0) = 0$. The α was chosen so that function (2.45) would become $1/R$ as rapidly as possible and would have no points of inflection. Thus

$$\begin{aligned} &\alpha = 2, \ a_0 = 1, \quad a_1 = 0.9760940, \\ &a_2 = -0.0478120, \ a_3 = 2.0340294, \\ &a_4 = -1.3490009, \ a_5 = 0.3116263. \end{aligned} \tag{2.45'}$$

Here, the unit of measurement of R is 1.40000 Å, the bond length in benzene. The table below lists values recommended by Pariser for $\gamma(R)$ at distances encountered in unbranched polyacenes.

Table 1

R in Å	$\gamma(R)$ in eV	R in Å	$\gamma(R)$ in eV	R in Å	$\gamma(R)$ in eV
0.0	10.959	5.560	2.563	8.680	1.652
1.390	6.895	6.059	2.355	9.360	1.490
2.407	5.682	6.370	2.242	9.730	1.475
2.780	4.978	7.223	1.981	10.023	1.432
3.678	3.824	7.355	1.946	10.856	1.323
4.170	3.390	7.739	1.850	11.033	1.302
4.812	2.949	8.455	1.695	12.038	1.193
5.012	2.836				

The differences between the results of calculations using (2.45) and those given in this table are unimportant, and lie within the limits of error due to interpretation of the experimental results on the spectrum of benzene. (Similarly, Ruedenberg's inclusion of magnitudes of the order of s^2 make no appreciable difference in the final result).

The spectrally-determined interval between the excited states of benzene and the ground state allows us to determine the resonance integral. Its values are given in Table 2:

Table 2

$-\beta$ in eV	Reference	Remarks
2.371	[40]	Determined from the spectrum $s = 0$
2.389	[37]	Determined from the spectrum, including s^2 in the calculation
2.13	[33]	Determined from the heat of combustion at $\gamma \sim 1/R$

Calculating by this method the values of the resonance integrals in various instances, i.e., at various R, and assuming the resonance integral to be an exponential function of the interval, we can find the parameters of this exponential function. Thus, in the hydrocarbons [29]

$$\beta(R) = -6442 \exp(-5.6864\ R)\ \text{eV}\ (R\ \text{in Å}). \tag{2.46}$$

Similar calculations have been carried out for molecules containing nitrogen, but by a simpler method which uses second-order interpolation of $\gamma(R)$ [29, 41]. The data for the parameters are given in Tables 3 and 4.

For heteroatoms, such as nitrogen, it is also necessary to introduce an empirical correction δW for the difference in the electronegativities of nitrogen and carbon. The best result for the nitrogen-containing heterocycles are obtained when $\delta W = -1.68$ electron volts

[17, 18], which is very close to the real difference between the electronegativities of the two atoms.

Table 3

R in Å	$\gamma_{NN'}$ [29]	$\gamma_{NN'}$ [41]
0.0	12.27	12.833
1.36	8.00	—
2.36	5.72	—
2.75	5.03	5.156

Table 4

R in Å	$\gamma_{CN'}$ [29]	$\gamma_{CN'}$ [41]
1.36	7.68	7.977
2.38	5.60	5.725
2.76	4.97	—

$$\beta_{CN} = -2.576 \text{ eV [29]}, \quad \beta_{CN} = -2.74 \text{ eV [41]}.$$

The results quoted above indicate that, in spite of differences in the method of determination, the empirical methods lead to practically the same values of all parameters. The slight discrepancies, of the order of 0.1 electron volts, may possibly serve as an indication of the accuracy of the empirical method.

The most important deviation of the empirical parameters from their calculated values is the decrease in $\gamma(R)$ at small R. The nature of this decrease is not difficult to understand. The integrand expression of $\gamma(R)$ is proportional to the product of probabilities $|\chi_p|^2 \cdot |\chi_q|^2$ of finding two electrons at small distances to each other (at small R_{pq}). The wave function of the MO method does not account for the correlated motion of the electrons, i.e., the mutual influence of their motions, which, as a result of Coulomb repulsion, produces a tendency of the electrons to be far from each other. Thus, for small R_{pq}, the probability value will be overestimated in the MO method. Consequently, the decrease in the values $\gamma(R)$ at small R take correlation indirectly into account. This argument, together with the many satisfactory results of calculations based on the above method allow us to conclude that, in the above method of bringing in empirical parameters, the latter sufficiently "absorb" all inaccuracies of the scheme. Any further increase in the accuracy would apparently require a different calculation procedure.

There are some approaches which take correlation into account and attempt to improve the self-consistent method. In the paper of Julg [39], correlation is taken into account by the introduction of factors $f(r_{pq})$ which depend on the interelectronic distances and which should become zero at $r_{pq} = 0$. In the final analysis, it is found that $f(r_{pq})$ appears in the integrand expression for $\gamma(R)$, causing a reduction of this function at small R.

The same result is obtained by the method of "partitioned p-orbitals," worked out by Dewar [43]. It is assumed that electrons having spin α occupy MO's distributed above the plane of the molecule, while electrons having spin β are in the symmetric MO's lying below this plane. This reflects the tendency of electrons with opposite spins to be far from each other. The MO's of electrons with spin α are expanded over the upper normalized "fractions" of the dumbbell-shaped p-orbitals, and the MO's of electrons with spin β are expanded over the lower "fractions" (with the same coefficients). The basic equations for these coefficients retain the form of (1.61), but certain integrals of the electron interaction between $2p_z$-orbitals in the energy matrix are replaced by analogous integrals between "fractions" and are very close to the empirical values (the one-electron integrals over the "fractions" are equal to the integrals over the "total" orbitals). However, the total wave function used by Dewar is not an eigenfunction of the square of the spin and has a number of other defects [44].

Therefore, both these approaches lead merely to modifications in the equations for the parameters, improving the agreement between calculated and empirical values. The whole method of calculation remains, however, unaltered.

A slightly different technique is the so-called electronegativity variation method due to Brown [45]. In the most widely used procedure for finding the molecular wave functions by the LCAO MO method (described above), one starts from fixed exponents of the atomic functions. However, as has been stated in the first section, the orbital exponents should, in principle, be determined simultaneously with the variation of the exponents. Brown's method is a semi-empirical variant of this procedure. Rather than determine the absolute minimum of the energy as a function of exponents and coefficients $E=E(\zeta, C)$, it approaches the minimum; this approach is not, as is usual, over the "plane" $\zeta = \text{const}$, but over a certain "surface" $\zeta=\zeta(C)=\zeta(P)$. The dependence of the orbital exponents on elements of the bond-order matrix (equation of the "surface") is derived by Brown by extending Slater's semi-empirical rules for finding the effective charges to the case of a non-integral number of electrons in the atom. For the same reason, for which the effective Slater charges in the atom are "better" than fixed charges, we must conclude that $\zeta = \zeta(P)$ passes closer to the absolute minimum of $E=E(\zeta, P)$ than $\zeta = \text{const}$. The relation of ζ_q with the charge on this atom was obtained on the basis of Slater's formula

$$Z_q=N_q-1{,}35-0{,}35(\sigma_q+P_{qq})=Z_q^0-0{,}35\Delta P_{qq}, \qquad (2.47)$$

where ΔP_{qq} is the deviation from unity of the π-electron charge on the qth atom. The factor Z_q appears in the formulae for all integrals containing atomic functions. In reality, however, Brown considered only the dependence of the Z_q on the ionization potential (second-order interpolation) and of the once-center integral γ_0 (linear interpolation).

The method was most successful with compounds containing heteroatoms and with non-alternant hydrocarbons of non-uniform charge distribution. The methods smooths out the electron density so that the results show a better correspondence to the experimental data.

In connection with the above, we must say that the question of convergence of the various iteration methods of solving our non-linear equations is still a completely virgin territory from the mathematical point of view, even in the case of the most widely used method of self-consistency of bond orders. This is even more true of the method of self-consistency of electronegativities or bond lengths. This is surprising since this problem is of current interest in view of the fact that the solutions of these equations are not unique, and the iteration process does not necessarily yield the required solution. In fact, Eqs. (1.56) are not linear but cubic. This problem would be best studied on the example of the bond order equations (1.61).

Even assuming that F_0 is independent of P, the second group of equations in (1.61) does not determine P uniquely but gives $m!/n!(m-n)!$ solutions [46], corresponding to all states resulting from excitation of pairs of electrons. The solution giving the minimum energy corresponds to the ground state. This ambiguity is natural, since the basic equations have been obtained from the condition that the first variation is equal to zero, the necessary though insufficient condition for a minimum.

The introduction of a dependence of F_0 on P (self-consistency) can lead to an additional non-trivial ambiguity as a result of an increase in order of the equations.

Let us consider, for example, the molecule of butadiene, which has also been studied by Hall [47]. Since the present author believes that Hall's results may be open to doubt, and because this example throws light on all aspects of the problem, let us examine this case in some detail.

The molecule of *trans*-butadiene can be characterized by parameters β_{12}, β_{23}, γ_{12}, γ_{23}, and γ_{14}. The solution for the bond orders will be sought in the form characteristic of alternant hydrocarbons (cf. Section III).

$$P = \frac{1}{2}\begin{pmatrix} I & P_{*0} \\ P'_{*0} & I \end{pmatrix}, \tag{*}$$

where

$$P_{*0} = \begin{pmatrix} P_{12} & P_{14} \\ P_{23} & P_{34} \end{pmatrix} = \begin{pmatrix} x & y \\ z & x \end{pmatrix}.$$

Expression (1.61′′, 2) then gives three equations

$$x^2 + y^2 = 1,\ x^2 + z^2 = 1,\ x(y + z) = 0,$$

and (1.61′′, 1) reduces the condition that $F_{*0}P'_{*0}$ be symmetric, a condition in which

$$F_{*0} = \begin{pmatrix} \beta_{12} - \frac{x}{2}\cdot\gamma_{12} & -\frac{y}{2}\gamma_{14} \\ \beta_{23} - \frac{z}{2}\cdot\gamma_{23} & \beta_{12} - \frac{x}{2}\gamma_{12} \end{pmatrix},$$

This equation may be written in the form

$$\beta_{12}z - \frac{\gamma_{12}}{2}xz - \frac{\gamma_{14}}{2}xy = \beta_{23}x - \frac{\gamma_{23}}{2}xz + \beta_{12}y - \frac{\gamma_{12}}{2}xy.$$

Two solutions of the first type, $y = z = 1/2$ and $y = z = -1/2$ are obtained at $x = 0$. For $y = -z$ the first two equations become identical, $x^2 + y^2 = 1$, and the symmetry condition reduces to $\mu xy + \lambda y = x$, where $\lambda = -2\beta_{12}/\beta_{23}$ and $\mu = (2\gamma_{12} - \gamma_{14} - \gamma_{23})/2\beta_{23}$, and where the parameter μ is the interaction factor. In the absence of interaction ($\mu = 0$), we thus obtain two solutions, corresponding to the points of intersection of the circle by the straight line $\lambda y = x$. There will thus be six solutions in all [there are two other solutions which can't be represented in form (*)] as, in fact, follows from the above considerations. At $\mu \neq 0$, one set of values of y and z corresponds to each value of x. Hall attempted to distinguish two sets, differing in the signs of x and y, but, in fact, his second equation reduces to the first when α is replaced by $\pi - \alpha$. Actually new solutions, in which the interaction can appear, may be possible if the second arm of the hyperbola $\mu xy + \lambda y = x$ is made to intersect the circle. From the condition of tangency of the hyperbola to the circle, $1 - \mu y/\mu x + \lambda = -x/y$, we obtain an inequality which determines the possibility of the existence of these additional solutions.

$$\mu^{2/3} \geqslant \lambda^{2/3} + 1.$$

This inequality is not satisfied at values of parameters assumed by Hall, although new solutions could appear at strong interactions $\gamma \gtrsim 6|\beta|$.

THE SIMPLE LCAO METHOD

1. INTRODUCTION OF PARAMETERS

From the practical point of view, the main shortcoming of the self-consistency method is the cumbersome iteration solution of the equations involved. For this reason, in the early applications of the MO method to the theory of molecules, a simplified approach, suggested by Hückel [49] and based on another method of introducing parameters, was used almost exclusively. This approach is known as the simple LCAO method.

If we neglect the dependence of the matrix elements F on the nature of the electron distribution within the molecule, i.e., if we neglect the dependence on R, then an element F_{pq}, represented in the form $\int \chi_p F_c \chi_q dv$, is determined only by the mutual alignment of atoms p and q, and does not depend on their position in the molecule. The F_{pq} themselves may then be regarded as empirical parameters characteristic of each pair of atoms.

Only two types of parameters are generally used in the simple LCAO method:

$$F_{qq} = \alpha_q^{\text{emp.}} \tag{3.1}$$

$F_{qt} = \beta_{qt}^{\text{emp.}}$ (q and t being neighboring atoms), while in all other cases $F_{qt} = 0$.

These assumptions are carried further, and we assume that the values of the parameters for each bond (in this method, the bond type depends on its chemical nature and sometimes on its length) are constant for a whole group of molecules of similar structure.

The choice of parameters is of critical importance. It is not unique, and depends on the property of the molecule (for which we want the best agreement with experiment). For example, in hydrocarbons, $\beta^{emp.} \sim -1$ electron volts is generally used to determine the energy of formation, and $\beta^{emp.} \approx -3$ electron volts to calculate the ionization potentials and the spectra (α disappears from most calculations).

A particularly large spread in the values of the parameters used in this method is encountered in the presence of heteroatoms. The electron density distribution is determined mainly by the parameters δ and ρ which characterize the deviation of the Coulomb integrals from their values in hydrocarbons:

$$\alpha_h = \alpha^{emp.} + \delta\beta^{emp.}, \quad \alpha_C = \alpha^{emp.} + \rho\beta^{emp.}, \tag{3.2}$$

where α_h is the Coulomb integral for the heteroatom and α_C the Coulomb integral for its adjacent carbon. For example, in the case of nitrogen, the values of δ given by various authors ranged from 2 to 0.2. The most reliable values for the six-membered heterocycles are probably the $\delta = 0.20$, $\rho = 0$, recommended by McWeeny [17]. With this value of δ, Hückel's method gives practically the same electron density distribution as the self-consistency method.

The dependence of β_{tq} on bond length R_{tq} is usually assumed to be an exponential function:

$$\beta_{tq} = \frac{\beta^{emp.}}{S} S_{tq} = \beta_0 \exp(-\alpha R_{tq}). \tag{3.3}$$

In the case of hydrocarbons, this deviation of β_{tq} from $\beta_0^{emp.}$ is usually only taken into account in calculations of internuclear distances by the "self-consistent bond-length" method. This method is based on a calculation of the dependence of the lengths of C-C bonds on their order. In place of the complex formulations of this dependence used previously [50-52], the linear expression [53]

$$R_{tq} = 1.517 - 0.180 P_{tq} \text{ Å} \tag{3.4}$$

is most frequently used today.

Thus, after the bond orders are found, the initially assumed bond lengths should be revised using (3.4). This leads to changes in the elements of the energy matrix in accordance with (3.3) (in hydrocarbons $\alpha = 2.683 \text{Å}^{-1}$). It follows that a new determination of bond orders then becomes necessary, and the process is repeated until "self-consistency" is achieved. Since in conjugated hydrocarbons the deviation of bond lengths from the equilibrium length of 1.397Å found in benzene is negligible, the above calculation can be carried out by the perturbation theory. Coulson and Golebiewski [53] showed that this leads to a system of linear equations for direct determination of the corrections to bond lengths. The bond lengths found by the above authors for naphthalene, butadiene, and anthracene were in excellent agreement with experiment.

If, however, we do not aim at a determination of bond lengths, we can consider them all to be equal, in which case all β_{tq} are the same. In this "simplest LCAO method," we have for conjugated hydrocarbons

$$F^0 = \alpha^{\text{emp.}} I + \beta^{\text{emp.}} M. \tag{3.5}$$

Formally, this expression is obtained if the integral term in (2.39) is neglected. It is nevertheless assumed that the interaction of π-electrons is "effectively" taken into account in the simple LCAO method, by changes in the parameters $\alpha^{\text{emp.}}$ and $\beta^{\text{emp.}}$ in relation to the α and β in (2.32).

The assumption that the energy matrix has the form (3.5) leads to a number of consequences which are important from the point of view of the chemistry of conjugated molecules. To set up the topological matrix M it is sufficient to know only the structural formula of the given molecule. This means that all properties of the molecule (which can be calculated from wave functions, knowing only F^0) are determined by its chemical structure, i.e., by the sequence of chemical bonds present in the molecule. Such a treatment corresponds to Butlerov's theory of chemical structure. In addition, the simple LCAO method takes into account the mutual influence of atoms in the molecule. This is because all properties, including local properties at any two given atoms (e.g., bond strength, which is determined by the bond order) are characterized by the overall matrix M, the arrangement of all the atoms in the molecule, and the molecular structure. All of this explains the success of the simple LCAO theory, which to some extent reflects the objective data of organic chemistry.

The simplicity of Hückel's method opens wide possibilities for its application to the study of various properties of conjugated molecules. At the present time, only the very large hydrocarbon molecules remain to be studied by this method. The satisfactory agreement of the results with the experiment has led to attempts at constructing a more rigorous foundation for the simple LCAO method.

2. THE METHOD OF EQUIVALENT ORBITALS

The main problem of the LCAO method is the diagonalization of the matrix whose elements are empirical characteristics of individual structural formations of the molecules. This problem can also be approached from a somewhat different point of view.

According to (1.36), the unitary transformation of the MO's in the Hartree-Fock equations (1.34) amounts to putting the matrix Λ of energy parameters in diagonal form

$$\Lambda = \int \omega^{+} f \omega dv, \tag{3.6}$$

where f is Fock's operator. Its elements are given by equations analogous to those used for F_{tq} (1.47):

$$\Lambda_{tq} = \int \bar{\omega}_t f \omega_q dv. \tag{3.6'}$$

The difference between the expansions of the MO φ in atomic functions χ and in functions ω is that the number of the ω functions is equal to the number of the φ functions. We can thus limit ourselves to n functions of the expansion because the ω functions satisfy (1.36), whereas the χ functions are arbitrary. The problem of finding the φ from the ω (i.e., diagonalization of Λ) is equivalent to the problem of solving the exact Fock equations given by (1.32). However, the ω remain unknown, just as the φ. The question thus arises as to whether the functions ω could not be defined in such a way that the elements of the matrix Λ could be regarded as empirical parameters related to definite structural elements of the molecule.

The answer to this may be obtained by generalizing the following properties of symmetrical molecules. Since the matrix ε is diagonal in such molecules, the MO's φ should, according to (1.37), form a basis of irreducible representations of the symmetry groups of the molecule. They are therefore different from zero within the entire volume of the molecule, and describe the collective properties of the molecule, such as excitation and ionization. The matrix Λ is non-diagonal, and the functions ω transform under symmetry operations as reducible representations. They can be chosen in such a way that each ω_k is related to a certain symmetrical region of the molecule (for example, a bond or an atom) and during symmetry operations the ω_k then transform into each other. In this sense, they are equivalent to each other, and are consequently known as equivalent orbitals. In other words, the equivalent orbitals are localized in various symmetrical regions of the molecule, and resemble AO's in this respect. The characteristics of the equivalent orbitals are determined only by their immediate environment.

For this reason, a system of equivalent orbitals ω can be constructed in terms of the individual locally equivalent regions (identical atoms, bonds) even in the absence of symmetry. The Λ_{pq} can thus be regarded as identical for the identical structural elements; this relationship holds not only for any one single molecule but also for all molecules of similar structure. The factor Λ_{pq} can thus be assumed to be an empirical parameter.

The above reasoning, which is the essence of the method of equivalent orbitals [54, 55], is applicable only when the number of locally equivalent regions of the molecule is equal to the number of MO's and to half the number of electrons. Therefore, the method cannot be applied to the π-electrons, because the number of π-electrons is equal to the number of carbon atoms, not half of this number. On the other hand, the conditions for the applicability of the method are satisfied in the case of saturated compounds, in which each bond is composed of a pair of electrons. Thus, the LCAO method was justified not for the π- but for the σ-bonds!

In fact, the increasing use of the simple semi-empirical LCAO method in calculation of σ-bonds promises far-reaching results. This method allows us to calculate such fine effects as changes in the ionization potentials [56, 57] and the heats of combustion [58] within homologous series, and provides a satisfactory description of the electron-density distribution and reactivity [59]. The basic results depend only slightly on details of the calculation. The parameter $\beta^{emp.}$ is about -6.4 electron volts in the calculations of the ionization potential, and about -1.7 electron volts in the case of the heat of combustion. The second parameter, determining the results of the calculation, characterizes the resonance integral between σ-orbitals originating at the same atom. It is equal to 0.34–0.35 in units of $\beta^{emp.}$. The LCAO method is therefore just as applicable to saturated as to conjugated compounds.

To provide an analogous theoretical basis for the use of the LCAO method in conjugated molecules, Hall proposed the method of the "standard excited state" [60]. He considered a strongly excited state, in which all π-electrons have parallel spins. Then the number of space one-electron functions—the so-called standard MO's in the Hartree-Fock equations—is equal to the number of π-electrons, i.e., the number of atoms in the conjugated system. Thus, there is a possibility of expressing the relationships in terms of "standard equivalent orbitals" localized around the carbon atoms. The basic assumption of the method of "standard excited state" is that it is permissible to neglect the changes occurring in the MO's upon excitation and to assume that the ground state MO's are equal to one-half of the standard MO's which correspond to the lowest orbital energies. Each such orbital contains two π-electrons. However, direct calculations using this "standard excited state" method [61] lead to less satisfactory results than do a number of other methods. It seems that this state is too strongly excited for the MO's to be regarded as "nonpolarized."

A more direct explanation of the success of the simple LCAO method can be given on the basis of the similarity of its equations for the MO's with the equations for self-consistent MO's in the ground, rather than in the excited state of the molecule.

3. SPECIAL FEATURES OF ALTERNANT HYDROCARBONS

The so-called alternant hydrocarbons, whose special characteristics were first studied by Coulson [62], occupies a special position among conjugated molecules. The alternant hydrocarbons are distinguished by the fact that the carbon atoms participating in the conjugation can be divided into two sets such that the atoms in one class will not be vicinal to others from their own set. These compounds include the polyenes, polynuclear polycyclic hydrocarbons composed of benzene rings, as well as others. However, molecules containing the cyclopentadiene group (such as azulene, fluoranthene, decacycline, etc.) are not alternant.

Mathematically speaking, the characteristic property of alternant hydrocarbons is that the matrices H_{eff}, M, and others, related to short-range forces, can, by suitable numeration of the atoms, be expressed in block form in the "tight-bonding" approximation. For example:

$$M = \begin{pmatrix} 0 & B \\ B' & 0 \end{pmatrix}. \tag{3.7}$$

In this notation the first atoms refer to atoms of the first set (*) and the following to the atoms of the second set (0).

The characteristics of the bond-order matrix for the simple LCAO method in approximation (3.5) can be determined from formula (1.62), taking account of the fact that operations on block matrices can be carried out by treating the blocks as the usual elements but preserving the order of the multipliers. Consequently,

$$P = \frac{1}{2} \begin{pmatrix} I & B\,(B'B)^{-1/2} \\ B'\,(BB')^{-\frac{1}{2}} & I \end{pmatrix}. \tag{3.8}$$

where primes indicate transposes. Equation (3.8) leads to some results of great physical importance. Namely, the result that all charges on atoms $2P_{ss}$ are equal to unity (uniform charge distribution), and all orders of bonds between atoms of the same set are equal to zero. The orders of bonds between atoms of different sets can be calculated from Hall's formula

$$P_{*0}=B(B'B)^{-1/2}. \tag{3.9}$$

Alternant hydrocarbons also have special distributions of energy levels. In accordance with (3.5), the eigenvalues of F_0 are

$$\varepsilon_i=\alpha^{\text{emp.}}+\beta^{\text{emp.}}\cdot x_i, \quad i=1, 2\ldots, m_*+m_0, \tag{3.10}$$

where the x_i are eigenvalues of the matrix M.

$$M^2=\begin{pmatrix}BB' & 0\\ 0 & B'B\end{pmatrix}. \tag{3.11}$$

where the diagonal block BB' has dimension equal to the number of atoms in the first set m_*, dim $(B'B)$ equals the number of atoms in the second set m_0. If, for example, $m_*\leqslant m_0$, this automatically includes radicals and biradicals of the type found in triangulene and in Chichibabin's hydrocarbon. M^2 has m_* eigenvalues, which can be determined from

$$|BB'-x_i^2I|=0. \tag{3.12}$$

These eigenvalues are doubly degenerate since $B'B$ also has the same eigenvalues, and moreover, the eigenvalue zero is (m_0-m_*)-fold degenerate because the rank of $B'B$ is m_* and its dimension is m_0. The energy levels are therefore distributed symmetrically with respect to the (m_0-m_*)-fold degenerate value of $\alpha^{\text{emp.}}$. Each level in the lower half is occupied by two electrons, and the level $\alpha^{\text{emp.}}$ contains $m_0+m_*-2m_*=m_0-m_*$ π-electrons which, according to Hund's rules, should have parallel spins. It follows that the theory of open shells must be applied to systems in which $m_*\neq m_0$.

M^2 is reduced to diagonal form by unitary transformation with the aid of a matrix Q,

$$Q=\begin{pmatrix} C^* & 0 & \\ 0 & C^0 & C^{\dagger} \\ \underbrace{}_{m_*} & \underbrace{}_{m_*} & \end{pmatrix}\begin{matrix} \} m_* \\ \} m_0, \\ \\ \end{matrix} \tag{3.13}$$

where, in the second diagonal block, we have selected all the eigenvectors $C^{\dagger}$ which corresponds to the eigenvalue zero. The corresponding eigenvectors of M are obtained from those of M^2 by transformations in m_* two-dimensional spaces which combine the ith and $(i+m_*)$-th columns of Q, and a transformation in (m_0-m_*)-dimensional space which combines the last (m_0-m_*) columns among themselves. Therefore, the matrix which brings M to diagonal form can be written in the block form

$$V=\begin{pmatrix} C^* & C^* & 0 \\ C^0 & -C^0 & C^{\dagger} \end{pmatrix}. \tag{3.14}$$

The blocks of this matrix must obey the conditions

$$BC^0=C^*X,\ B'C^*=C^0X,\ BC^{\dagger}=0, \tag{3.15}$$

where X is a diagonal matrix $[x_1, x_2, \ldots, x_{m_*}]$. The last of these three equations determines the so-called non-bonding MO's. It has non-zero solution because not all the columns of B are linearly independent. There are at least m_0-m_* dependent columns, and hence, the same number of non-bonding MO's. The presence of a zero block in V means that electrons with unpaired spin are distributed only among the atoms of the "large" (second) set [64].

All these results (at $m_*=m_0$) also remain valid in the self-consistent method with the energy matrix (2.39). Now, looking for solution for the bond orders in the form

$$P=\frac{1}{2}\begin{pmatrix} I & P_{*0} \\ P'_{*0} & I \end{pmatrix}, \tag{3.16}$$

we note that F^0 also retains block form

$$F^0 = \begin{pmatrix} -WI & F_{*0} \\ F'_{*0} & -WI \end{pmatrix}, \tag{3.17}$$

where W is the electronegativity of carbon, while F_{*0} has elements which can be determined from (2.41) and (2.42). This form of F^0 shows that (3.16) has a solution. But now

$$P_{*0} = F_{*0}(F'_{*0}F_{*0})^{-1/2}. \tag{3.18}$$

This agreement between the basic results of the self-consistent and simple methods confirms the validity of the latter in the case of alternant hydrocarbons. The results obtained on applying the simple method to the electronic structure of the ground state of alternant hydrocarbons are also in good agreement with experimental data. If we compare the expressions for the total energies and the energy matrices as given by the self-consistent and the simple method and neglect in the latter the long-range forces [i.e., assume that $\gamma(R) = 0$ for R greater than the length of one bond], we can obtain two sets of parameters of the simple method useful for calorimetric and spectroscopic calculations [63].

The simple LCAO method can thus successfully replace the more complicated theory in the study of a whole range of problems involving alternant hydrocarbons. The method represents a crude but reliable theory of molecular phenomena, and has been amply confirmed in practice.

It must be stressed that the results of the self-consistent and the simple method differ considerably in the case of non-alternant hydrocarbons, with the simple LCAO approach giving poor agreement with experimental values.

This discussion of the simple LCAO method concludes the present review, in which we purposely made no reference to the calculation of systems containing an odd number of electrons. Such calculations form a separate large field for application of the MO method.

1. V. A. Fock. Trudy Gosud. Opt. Inst., 1931, 5, No. 51.
2. C. C. J. Roothaan. Revs. Mod. Phys., 1951, 23, 69.
3. L. C. Allen and A. M. Karo. Revs. Mod. Phys., 1960, 32, 275.
4. D. Hartree. Raschety atomnykh struktur [Russian translation of "Calculation of Atomic Structures"] State Press for Foreign Literature, 1960.
5. V. A. Fock. Yubileinyi sbornik. Akad. Nauk SSSR, 1948 (The Anniversary Volume of Acad. Sci. USSR, 1948).
6. A. T. Amos. Mol. Phys., 1962, 5, 91.
7. T. Koopmans. Physica, 1933, 1, 104.
8. R. S. Mulliken. J. Chem. Phys., 1955, 23, 1833.
9. K. Ruedenberg and N. S. Ham. J. Chem. Phys., 1958, 29, 1215.
10. B. Ransil. Revs. Mod. Phys., 1960, 32, 239, 245.
11. R. K. Nesbet. Revs. Mod. Phys., 1960, 32, 272.
12. D. A. McLean. J. Chem. Phys., 1960, 32, 1595.
13. R. McWeeny. Proc. Roy. Soc., 1956, A235, 496 and A237, 355; 1957, A241, 239.
14. P. O. Löwdin. J. Chem. Phys., 1950, 18, 365.
15. M. M. Mestechkin. Vestnik Leningrad Gosud. Univ., 1960, 22, 89.
16. I. S. Berezin and N. P. Zhidkov. Metody vychislenii [Methods of Calculation]. Fizmatgiz, 1959.
17. R. McWeeny and T. E. Reaczock. Proc. Phys. Soc., 1957, A70, 41.
18. A. T. Amos and G. G. Hall. Mol. Phys., 1961, 4, 25.
19. K. Ruedenberg, C. C. J. Roothaan and W. Jauhzemis. J. Chem. Phys., 1956, 24, 201.
20. C. C. J. Roothaan. J. Chem. Phys., 1951, 19, 1445.
21. K. Ruedenberg. J. Chem. Phys., 1951, 19, 1459.
22. I. Tauber. J. Chem. Phys., 1958, 29, 300.
23. R. S. Mulliken. Revs. Mod. Phys., 1960, 32, 232.
24. E. Clementi and A. D. McLean. J. Chem. Phys., 1962, 36, 745.
25. S. R. Boys. Proc. Roy. Soc., 1950, A200, 542.
26. I. C. Browne and R. D. Poshusta. J. Chem. Phys., 1962, 36, 1938.
27. K. Fukui et al. Bull. Chem. Soc. Jap., 1962, 35, 38.
28. R. S. Mulliken. J. Chem. Phys., 1949, 46, 497.
29. R. Pariser and R. G. Parr. J. Chem. Phys., 1953, 21, 767.
30. K. Ruedenberg. J. Chem. Phys., 1961, 34, 1861.
31. R. S. Mulliken. J. Chem. Phys., 1934, 2, 782.
32. K. Ruedenberg. J. Chem. Phys., 1961, 34, 1907.
33. I. A. Pople. Trans. Far. Soc., 1953, 49, 573.
34. G. G. Hall. Trans. Far. Soc., 1957, 53, 573.
35. K. Ruedenberg. J. Chem. Phys., 1961, 34, 1876.
36. M. Goeppert-Mayer and A. L. Sklar. J. Chem. Phys., 1938, 6, 645.
37. K. Ruedenberg. J. Chem. Phys., 1961, 34, 1897.
38. N. Mataga et al. Bull. Chem. Soc. Jap., 1958, 31, 453.

39. A. Julg. J. Chim. Phys., 1960, 57, 19.
40. R. Pariser. J. Chem. Phys., 1956, 24, 250.
41. T. Anno. J. Chem. Phys., 1958, 29, 1161.
42. H. A. Skinner and Pridchard. Trans. Far. Soc., 1953, 49, 1254.
43. M. J. S. Dewar and N. L. Hojvat. J. Chem. Phys., 1961, 34, 1232.
44. J. S. Griffith. J. Chem. Phys., 1962, 36, 1689.
45. R. D. Brown and M. L. Hafferman. Trans. Far. Soc., 1958, 54, 757.
46. M. M. Mestechkin. Vestnik Leningrad Gosud. Univ., 1960, 4, 12.
47. G. G. Hall. J. Chem. Phys., 1960, 33, 953.
48. M. G. Veselov. Zh. Eksper. Teor. Fiziki, 1938, 8, 795.
49. E. Hückel. Zs. Phys., 1931, 70, 204.
50. C. A. Coulson. Proc. Roy. Soc., 1939, A169, 413.
51. H. D. Deas. Phil. Mag., 1955, 46, 670.
52. T. Anno et al. Bull. Chem. Soc. Jap., 1957, 30, 638.
53. C. A. Coulson and A. Golebiewski. Proc. Phys. Soc., 1961, 78, 1310.
54. J. E. Lennard-Jones. Proc. Roy. Soc., 1949, A198, 1.
55. J. E. Lennard-Jones and J. A. Pople. Proc. Roy. Soc., 1950, A202, 166.
56. G. G. Hall. Proc. Roy. Soc., 1951, A205, 541.
57. V. M. Baranov and T. K. Rebane. Optika i Spektroskopiya, 1960, 8, 268.
58. G. Klopman. Helv. Chim. Acta, 1962, 45, 711.
59. K. Fukui et al. Bull. Chem. Soc. Jap., 1961, 34, 442.
60. G. G. Hall. Proc. Roy. Soc., 1952, A213, 102.
61. R. McWeeny. Proc. Phys. Soc., 1957, A70, 593.
62. C. A. Coulson and S. Rushbrook. Proc. Cambr. Phil. Soc., 1940, 36, 193.
63. M. M. Mestechkin. Zh. Fiz. Khimii, 1961, 35, 431.
64. H. C. Longuet-Higgins. J. Chem. Phys., 1950, 18, 265.

Chapter III

APPLICATION OF THE MO METHOD TO THEORETICAL STUDY OF THE REACTIVITY OF MOLECULES CONTAINING CONJUGATED BONDS AND AROMATIC MOLECULES

I. F. Tupitsyn and M. N. Adamov

INTRODUCTION

Even before the advent of quantum mechanics, the theoretical investigation of chemical processes followed two main trends.

1) Formal Kinetic Theory.

The rates of chemical reactions were studied without taking into account the structures of the reacting molecules. The experimental data were generally analyzed in terms of the empirical Arrhenius equation which expresses the reaction rate constant k as a function of temperature:

$$k = k_0 e^{-E^{a}/RT}, \qquad (1)$$

where k_0 and E^{a} are constants, E^{a} being the activation energy of the reaction.

2) Classical Study of the Reactivity of Organic Compounds.

This approach, founded on the work of Butlerov, Markovnikov, and Menshutkin, aimed at a quantitative determination of the relationships existing between the location and interaction of atoms in molecules on the one hand, and the ease with which a chemical transformation takes place when the organic molecule is attacked by various reagents, on the other.

We shall now briefly outline the development of each of these approaches in the "post-quantum" period.

The advances of the formal kinetic approach have been due to the development of a statistical method for calculating the rates of elementary reactions known as the activated complex method [1]. In this method the rate of reaction is related to the molecular characteristics of both the starting substances and of the activated complex, so that the formal character of the earlier theory is largely abandoned. In principle, this approach allows a calculation of the rate constant. The main object of quantum mechanics then becomes the discovery of how the potential energy of the reacting system depends on the properties of all the constituent atoms [2]. However, the solution of this problem involves such mathematical difficulties that the potential energy surface cannot be calculated with an accuracy sufficient for the prediction of reaction rates.

The inability to calculate the absolute activation energies has stimulated the developments of methods using approximate quantum-mechanical estimates of differences in activation energies of a series of reactions involving similar compounds. In other words, in these new attempts at developing the theory, we are considering the reactivities of various molecules and of various locations within the same molecule from the viewpoint of chemical kinetics.* The inaccuracies inherent in the assumed expressions for energy are thus cancelled out to some extent.

Estimations of the activation energy become practical if, in place of the true transition complex, we consider its simplified model. The latter must, however, be a molecular system whose structure reflects the essential features of the structure of the true transition complex.

The model of a transition complex first proposed by Wheland for substitutions into the benzene ring [6–9] has found wide application in the theoretical study of the reactivity of conjugated and aromatic systems. According to Wheland, the structure of a transition complex differs from that of the starting molecule in that bonds taken up by the attacking atom cease to be part of the conjugated system, so that the region over which π-electrons can move freely is reduced. In Wheland's model, the attacked carbon forms four σ-bonds, one of these being the coupling to the attacking atom.

The calculation is greatly simplified by the assumption that the activation energies of two (or more) similar reactions differ only in the part related to the behavior of π-electrons. This treatment disregards changes in the vibrational components of the activation energy ΔE^{a}_{v}, the σ-electron components ΔE^{a}_{σ}, or of changes in the energy of the σ—π interaction $\Delta E^{a}_{\sigma\pi}$. In other words, it is assumed that $\Delta E^{a} \approx \Delta E^{a}_{\pi}$). The value of ΔE^{a}_{π} for the reaction is then compared with

*The reactivity of a given compound in a given reaction depends both on thermodynamic parameters of the reaction and on kinetic factors.

In the present review we shall not be concerned with quantum calculations of reactions in which the reactivity is determined by thermodynamic factors; this aspect of reactivity has already been considered in [3, 4].

To limit the problem further, we shall not consider the effects of steric factors, solvation, and hydrogen bonding.

This means that our quantum-mechanical treatment applies to cases dominated by changes in the activation energy; only small changes occur in the entropy factor, and these may be neglected when comparing reaction rates.

The validity of this approximation is discussed in [5].

the calculated $\Delta E_{\pi}^{\prime a}$ value for another reaction which involves a related compound (or which involves the same compound but another site on it). The comparison of ΔE_{π}^{a} values then gives the relative reactivity of the first compound. The quantum-mechanical treatment based on Wheland's model has been called the "localization approximation."

Just as the trends in the formal kinetic theory, the development of the classical theory of chemical reactivity has tended toward incorporation of the conslusions of the electronic theory of the structure of atoms and molecules. A number of more or less successful hypotheses has been formulated, with the aid of which it was possible to provide an electronic interpretation of the rules governing the reactivities and the atomic interactions in organic molecules.

The most valuable of the above hypotheses [10, 11] were: a) the concepts of nucleophilic, electrophilic, and radical reactivity, whereby all organic substitutions are classified according to whether the electron pair forming the new bond is provided by the attacking reagent, by the substrate, or by both; and b) the theory of "electron shifts" which includes such concepts as inductive and mesomeric effects of atoms.

All these hypotheses assume that the reactivity is related to the distribution of electron charges in the starting molecule. The transition complex does not enter into consideration. The above hypotheses account correctly for a large volume of experimental data on the reactivity of individual positions of conjugated and aromatic molecules, but, for all their importance and value, cannot quantitatively relate the reactivity of a substance to the parameters characterizing its electronic structure.

The quite understandable desire to use the theory of "electron shifts" as the quantitative basis has led to the development of an approximate method of treatment in which reactivity is estimated via quantum-mechanical quantities characterizing the distribution of π-electron density in an isolated molecule. This is the isolated molecule approximation [11-15].

This treatment of reactivity (just as Wheland's transition complex model) is based on the assumption that in studying relative reactivities in a series of molecules of the same type, we can neglect changes in σ-electron energies and in the $\sigma-\pi$ interactions. Instead of a maximum on the energy curve (corresponding to the transition complex), one considers a state in which the energy is only slightly higher than that at the initial state, where there was no interaction between the attacked and attacking molecules. It is further assumed that the smaller this energy increment, the lower the activation energy.

The energy changes from the initial state are assessed by means of the so-called indices of the electronic structure of molecules. These indices comprise π-electron charges on atoms, bond orders, as well as free (or bound) valencies. None of these can be measured directly.

The structure indices used in the various approximate methods of quantum chemistry (the valence diagram method, molecular orbitals method, etc.) depend on the specifics of the given method. Thus, the numerical values of one and the same index vary in different methods, but lead, nevertheless, to analogous conclusions regarding the distribution of π-electron densities on atoms (or in bonds) and the relative electron deficiencies of atoms [16].

Starting from the theoretically calculated π-electron charges and bond orders, it is now possible to calculate π-electron dipole moments and C-C bond lengths and arrive at values in good agreement with the experiment. Such agreement cannot be accidental and proves that the electronic structure indices calculated by approximate quantum-mechanical methods must, to some extent, reflect the π-electron distribution in orbitals of conjugated molecules. This agreement is particularly noticeable in substitution reactions (nitration, sulfonation, amination, halogenation, free radical substitution, etc.) and *ortho* and *para* additions to conjugated and aromatic molecules. We thus get a feeling that the starting premises of both the "localization" and the "isolated molecule" approaches form a sound basis for a theory of chemical reactivity.

SOME FACTS CONCERNING THE REACTIVITY INDICES IN THE MOLECULAR ORBITAL METHOD

In their application to concrete cases, both of the above variants of the quantum-mechanical treatment of reactivity are most often used in connection with the MO method. While many modifications of this method are employed in the study of reactivity, its simplest form, that is, linear combination of atomic orbitals (LCAO MO) [17-22], as well as the free electron method (the "metallic" model of the molecule) [23-26] have found the widest applications. In analysis of complex conjugated molecules, the above two methods offer greater possibilities than the valence diagram approach because they account much better for the delocalization of the π-electron system.

1. ELECTRONIC STRUCTURE INDICES IN THE SIMPLEST VARIANT OF THE LCAO MO METHOD

In the LCAO MO method, the molecular orbital ψ of a π-electron is synthesized via a linear combination of atomic orbitals φ_r

$$\psi = c_1\varphi_1 + c_2\varphi_2 + \ldots + c_n\varphi_n, \tag{2}$$

where n is the number of atoms forming the conjugated system.

Considered as a separate function, the term $c_r\varphi_r$ in (2) corresponds to localization of the π-electron in the shell of atom r. Linear combination of φ_r reflects the fact the π-electron of a conjugated molecule is in the field of n nuclei forming the stable structure of the molecule.

For example, consider the derivation of the molecular orbitals of π-electrons in benzene

$$\psi_1 = \frac{1}{\sqrt{6}}(\varphi_1 + \varphi_2 + \varphi_3 + \varphi_4 + \varphi_5 + \varphi_6),$$
$$\psi_2 = \frac{1}{2\sqrt{3}}(\varphi_1 + 2\varphi_2 + \varphi_3 - \varphi_4 + 2\varphi_5 - \varphi_6),$$
$$\psi_3 = \frac{1}{2}(\varphi_1 - \varphi_3 - \varphi_4 + \varphi_6).$$

In the ground state of the molecule, each molecular orbital ψ_1, ψ_2, and ψ_3 is occupied by two electrons. The functions ψ_j (j = 1, 2, 3 . . . is the number of the molecular orbital) are normalized,* so that

$$\int \psi_j^2 d\tau = 1. \qquad (3)$$

The charge density at a point P, due to the π-electron occupying orbital ψ_j, is $\psi_j^2(P)$. The orbital coefficients $c_{1,j}$, $c_{2,j}$, . . . , $c_{n,j}$ may be used to characterize the contributions of each π-electron to the π-charges on carbon atoms and to C–C bond orders. In point of fact, in the generally used approximation $\int \varphi_r\varphi_s d\tau = 0$, it is easy to obtain from Eqs. (2) and (3), the relation

$$c_{1,j}^2 + c_{2,j}^2 + \ldots + c_{n,j}^2 = 1. \qquad (4)$$

This is interpreted by assuming that the square of $c_{r,j}$ represents that part of π-electron charge at atom r which is contributed by the electron in the jth molecular orbital. The effective π-electron charge of the rth atom is then given [27] by:**

---

*We shall limit ourselves to cases in which the wave function can be regarded as real.

**The quantity q_r gives the effective charge in units of e the elementary electronic charge. The so-called pure or residual changes $1-q_r$ are often given in place of q_r.

$$q_r = \sum_j g_j c_{r,j}^2, \tag{5}$$

g_j being equal to 1 or 2, depending on whether the given orbital is occupied by one or by two electrons.

For benzene, the π-electron density on atom 1 is

$$q_1 = 2\left[\left(\frac{1}{\sqrt{6}}\right)^2 + \left(\frac{1}{2\sqrt{3}}\right)^2 + \left(\frac{1}{2}\right)^2\right] = 1,$$

and it is easy to prove that $q_1 = q_2 = \ldots = q_6 = 1$.

The order of a conjugated bond between neighboring atoms r and s is a measure of that fraction of the total π-electron charge which is associated with the given r-s. bond. The bond order is calculated by adding the products $g_j c_{r,j} c_{s,j}$ for all occupied molecular orbitals [28]:

$$p_{rs} = \sum_j g_j c_{r,j} c_{s,j}. \tag{6}$$

Equation (6) leads to correct integral bond orders for molecules with isolated π-bonds: it is one for ethylene and two for acetylene. In conjugated and aromatic molecules the π-bond orders have values intermediate between zero and unity.

Thus the bond order between atoms 1 and 2 in benzene is

$$p_{1,2} = 2\left[\left(\frac{1}{\sqrt{6}}\cdot\frac{1}{\sqrt{6}}\right) + \left(\frac{1}{2\sqrt{3}}\cdot\frac{1}{2\sqrt{3}}\right) + \left(\frac{1}{2}\cdot 0\right)\right] = \frac{2}{3} = 0{,}667.$$

and all the bond orders of all the other carbon-carbon bonds in the ring are the same, i.e., 0.667.

Without going deeply into the problem, we shall note that the theoretically calculated bond orders and the CC bond lengths $L_{r,s}$ are approximately related [17, 18, 28] by:

$$L_{r,s} = L_S - \frac{L_S - L_D}{1 + \frac{\sigma(1 - p_{rs})}{k \cdot p_{rs}}}, \tag{7}$$

where L_S is the CC bond length in ethane, L_D the CC bond length in ethylene, and σ and k are the respective force constants of the CC bonds in the above two compounds. Equation (7) frequently allows a prediction of bond length from a knowledge of bond order. To illustrate its usefulness, consider Table 1, listing calculated and experimental bond lengths in naphthalene and anthracene [29, 52].

Table 1

Compound	Bond	Bond length, Å	
		calc. [29]	expt. [52]
Naphthalene	1—2	1.38	1.365
	1—9	1.41	1.425
	2—3	1.40	1.404
	9—10	1.42	1.393
Anthracene	1—2	1.38	1.370
	1—11	1.42	1.423
	2—3	1.41	1.408
	11—12	1.43	1.436
	9—11	1.40	1.396

The concept of bond order allows a determination of free valence, a value describing the degree of unsaturation of a carbon atom in a conjugated molecule.

If the sum of the orders of π-bonds ending at the rth carbon is N_r, and the maximum possible value of the sum of the π-bond orders of a carbon is N_{max}, then the free valence F_r for the carbon [30] is:

$$F_r = N_{max} - N_r. \tag{8}$$

The MO method gives a physical meaning to the concept of unsaturated valency, showing that for a given configuration of atomic nuclei in a molecule containing π-bonds, N_{max} assumes a strictly defined value of $\sqrt{3} = 1.732$* [32].

*Sometimes different values of N_{max} for primary (1.732), secondary (1.53), and tertiary (1.26) carbon atoms [31] are used.

The upper limit of "bound valency" (N_{max} = 1.732) for conjugated organic molecules may be found, for example, by calculating the sum of the orders of three π-bonds which end at a central carbon C_{centr} in symmetrical trimethylenemethane $(CH_2)_3C$. In this hypothetical biradical, each π-bond represents the maximum possible overlap of the π-electron clouds of C_{centr} and of the carbon of the CH_2 group, since the $2p\pi$-electron of the latter does not participate in any π-bond other than the bond with C_{centr} [12].

The free valence F_r resembles, to some extent, Thiele's "partial valence." There is, in fact, a fairly close correlation between the stability of a chemical compound and values of F_r. If we arrange compounds of various stabilities in order of increasing F_r at the most reactive atom:

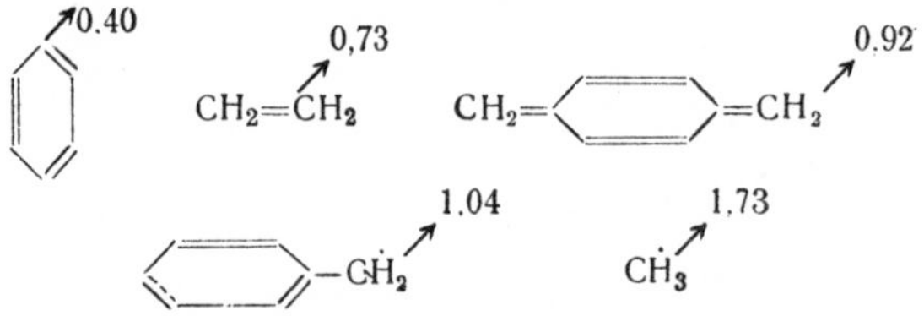

0.40 (benzene); $CH_2{=}CH_2$ 0,73; $CH_2{=}C_6H_4{=}CH_2$ 0.92; $C_6H_5{-}\dot{C}H_2$ 1.04; $\dot{C}H_3$ 1.73

we find that, beginning with *p*-dixylylene, all molecules are much too "spontaneously reactive" to exist as individual chemical species.

The quantities q_r, p_{rs}, and F_r form a set of electronic structure parameters in the LCAO MO method, and are conveniently represented as so-called molecular diagrams. For example, the molecular diagrams of naphthalene, pyridine, azulene, and butadiene are

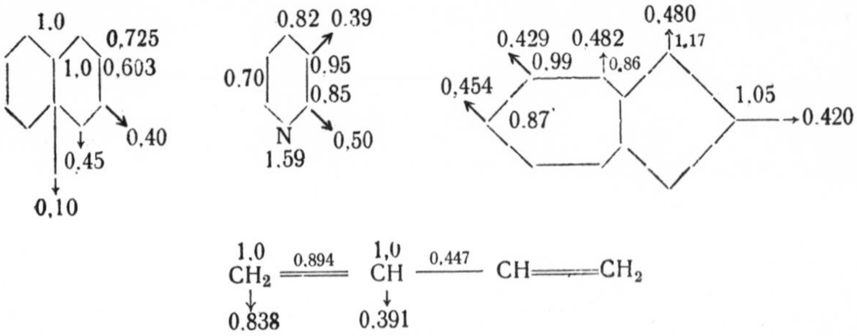

$$\underset{\downarrow\atop 0.838}{\overset{1.0}{CH_2}} \overset{0.894}{=\!=} \underset{\downarrow\atop 0.391}{\overset{1.0}{CH}} \overset{0.447}{-\!-} CH{=}CH_2$$

(The numbers of the atoms represent electron charges, those of the bonds stand for bond orders, and those at the ends of arrows denote free valencies).

2. ELECTRONIC STRUCTURE INDICES IN THE FREE ELECTRON METHOD

In the free electron method (that is, the simplest "metallic" model of a substance) it is assumed that the π-electrons move freely along certain lines corresponding to the contours of the molecule. Outside these lines, the potential energy of the electron is regarded as infinite. Electron interaction is neglected. The use of the "metallic" model is convenient because in this case the energy levels E_j of the π-electron are readily obtained, as well as the corresponding wave functions ψ_j, which become simple in this treatment. For example, in the case of polyene hydrocarbons:

$$E_j = \frac{h^2\pi^2}{2mL^2} j^2, \tag{9}$$

$$\psi_j = \sqrt{2/L} \sin \frac{j\pi x}{L}, \tag{10}$$

where L is the effective chain length of n conjugated bonds. $L = a(n + 2)$, where a is the interatomic distance.

The free electron method permits estimation of the π-electron density at any point x on the perimeter of the molecule

$$\omega(x) = \Sigma g_j \psi_j^2(x). \tag{11}$$

To characterize the π-electron charge distribution, the total π-charge is theoretically resolved into components by either of two alternative methods: charges on atoms $q(P)$ or on bonds $q(M)$ [33].

In calculating the π-electron charge on an atom P, the function $\omega(x)$ is integrated over a section l_p whose length is a. The center of this section should coincide with the center of the atom P

$$q(P) = \int_{l_p} \omega(x)\, dx. \tag{12}$$

To find the charge on bond $q(M)$, one integrates over a section l_M of length a, whose central point lies half-way along the bond between atoms P and Q.

$$q(M)=\int_{l_M}\omega(x)\,dx. \tag{13}$$

The degree of unsaturation of the carbon atoms in the "metallic" model is described by means of "occupied valencies" of atoms. The "occupied valencies" are defined as the sum of charges in all the bonds involving the carbon atom in question.

The molecular diagrams of naphthalene and azulene, obtained from calculations by the free electron method [33], are given below:

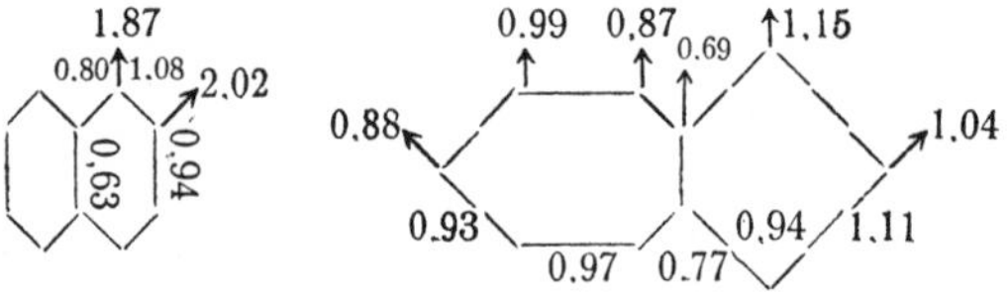

1.87
0.80 1.08 2.02
0.63 0.94

0.99 0.87 0.69 1.15
0.88 1.04
0.93 0.94 1.11
0.97 0.77

They show π-electron charges on bonds as well as on "occupied valencies" (at the end of arrows).

3. ORBITAL COEFFICIENTS, ENERGIES OF π-ELECTRONS, AND MUTUAL POLARIZABILITIES OF ATOMS AND BONDS IN THE LCAO MO METHOD

The possible energies ε_j of π-electrons occupying the molecular orbitals, as well as the coefficients $c_{r,j}$ of atomic orbitals φ_r, are calculated from the conditions $\partial\varepsilon_j/\partial c_{r,j}=0$. These conditions express the requirement of minimum energy (see, for example, [17, 18]). The values of $c_{r,j}$, minimizing the energy are then obtained as solutions of a system of n equations

$$(\alpha_r-\varepsilon_j)c_{r,j}+\sum_{s\neq r}^{n}\beta_{r,s}c_{s,j}=0, \quad r=1, 2, 3, \ldots, n. \tag{14}$$

where α_r is the so-called Coulomb integral of atom r and β_{rs} is the resonance integral of bond $r-s$. The integrals α_r and β_{rs} are not computed but are regarded as empirical parameters.

In the simplest variant of the LCAO MO method, it is assumed that the Coulomb integrals of all the carbon atoms of the conjugated system are equal to α, and the zero-point on the energy scale is so chosen as to make $\alpha = 0$. The Coulomb integrals of atoms other than carbon (in the conjugated system) are estimated from experimental values of the ionization potentials I of the atoms, and the difference $\alpha - \alpha_X$ (X is the heteroatom) is assumed equal to $I_C - I_X$. (Such estimate of α_X is, of course, only a rough approximation; for example, α_N for the nitrogen in pyrrole is, in fact, smaller than that for the nitrogen in pyridine; the nitrogen in pyrrole attracts π-electrons more strongly than the nitrogen of pyridine.)

The same value of β (usually 20 kcal/mole, from spectroscopic data) is assigned to all resonance integrals of carbon-carbon bonds. The integrals β_{rs} between non-adjacent atoms are disregarded.

The simplest variant of the LCAO MO method thus yields the following expression for the total energy of π-electrons of the molecule:

$$E_\pi = \sum_j g_j \varepsilon_j = \sum_r \alpha_r q_r + 2\sum_{r<s} \beta_{rs} p_{rs}. \tag{15}$$

Differentiation of (15) with respect to α_r and β_{rs} leads to the important relations

$$q_r = \frac{\partial E_\pi}{\partial \alpha_r}, \quad p_{rs} = \frac{1}{2}\frac{\partial E_\pi}{\partial \beta_{rs}}. \tag{16}$$

To describe the ability of a molecule to redistribute π-electron charges in response to attack by a variety of reagents, the LCAO

MO method uses quantities known as mutual atomic and bond polarizabilities [34]. These quantities are introduced as follows:

The quantity $\pi_{r,t}$:

$$\pi_{r,t} = \frac{\partial q_r}{\partial \alpha_t}, \tag{17}$$

which characterizes the rate of change of the π-electron charge at atom r in response to a change in the Coulomb integral of atom t in the same molecule, is called the mutual polarizability of atoms r and t. If $r = t$, we are dealing with the *self*-polarizability of atom r. The values of $\pi_{r,r}$ for several substances are given in Table 2. The mutual polarizability of atom t and bond rs, ($\pi_{rs,t}$), and the mutual polarizability of two bonds rs and tu, ($\pi_{rs,tu}$), are determined in an analogous manner:

$$\pi_{rs,t} = \frac{\partial p_{rs}}{\partial \alpha_t}, \tag{18}$$

$$\pi_{rs,tu} = \frac{\partial p_{rs}}{\partial \beta_{tu}}. \tag{19}$$

All polarizabilities can be represented as functions of orbital coefficients [34].

Polarizabilities may be used to estimate, for example, the changes in the π-electron charges at the atoms caused by placing a heteroatom or a substituent into the position r of the conjugated molecule. An incremental change $\delta\alpha_r$ in α_r causes the following change in the π-electron charge on atom s:

$$\delta q_s \approx \pi_{s,r}\delta\alpha_r. \tag{20}$$

4. ALTERNANT HYDROCARBONS

In studying reactivities by the LCAO MO method, it was found desirable to separate out a large class of hydrocarbons which are readily susceptible to theoretical calculations. These are the so-called alternant hydrocarbons, in which all carbons of the conjugated system may be placed in one of two groups such that no two neighboring atoms belong to the same group [35]:

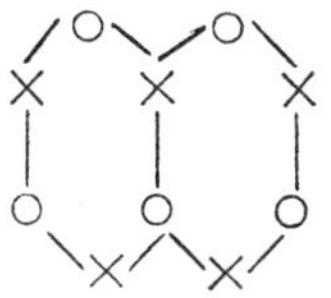

Such a division is shown above for naphthalene, the atoms belonging to different groups being denoted by crosses and circles, respectively. All aromatic hydrocarbons containing an even number of carbons in the ring as well as the polyenes (butadiene, hexatriene, and others) belong to this class.

The electronic structure of these compounds is distinguished by the fact [36] that:

(1) All carbons carry the same π-electron charges ($q_r = 1$).

(2) The self-polarizabilities of atoms are negative ($\pi_{r,r} < 0$). The mutual polarizabilities of atoms r and s assume alternately positive and negative values as the number of C–C bonds located between atoms r and s increases.*

(3) So-called anti-bonding molecular orbitals exist in all odd alternant hydrocarbons, (i.e., in those containing an odd number of carbons). The energy of such an orbital, $\varepsilon_0 = \alpha$, is independent of the resonance integral β (on the usual energy scale $\varepsilon_0 = 0$). The orbital coefficients $c_{s,0}$ of the anti-bonding MO for the group containing the smaller number of atoms (atoms marked with circles in the naphthalene diagram above) are equal to zero. The sum of $c_{r,0}$ for the other group (r denotes atoms marked with crosses) is $\sum_r c_{r,0} = 0$.

*From which follows, in accordance with (20), the rule of alternating polarity, well known in chemistry.

SUBSTITUTION REACTIONS

1. THE "ISOLATED MOLECULE" APPROXIMATION

As has already been mentioned, the starting point of the "isolated molecule" approximation is the assumption that the energy curves of two comparable reactions do not intersect between the initial point I and the transition complex III (Fig. 1). Comparison of the course of the two curves over a small section between I and II then allows a prediction of the relative positions of their maxima.

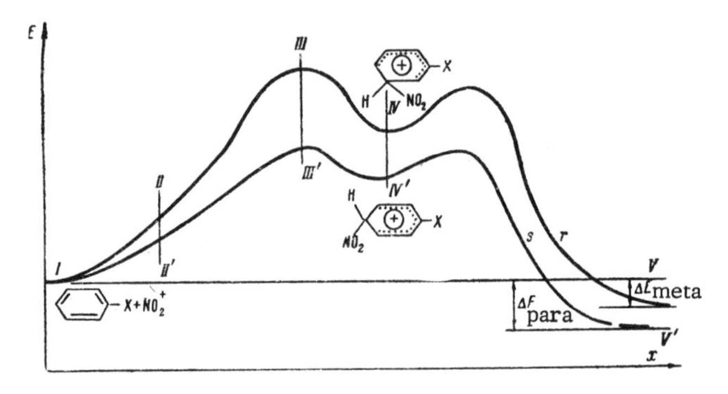

FIG. 1. Energy curves for the electrophilic substitution in monosubstituted benzene. x is the reaction coordinate.

Consider a small change of energy of the system when one of the carbons is attacked by an electrophilic agent.

We shall assume that the curves of Fig. 1 represent the energy profile of the nitration of singly substituted benzene C_6H_5X (X = OH, NH_2). These substituents are known to be *ortho-para* directing, with virtually no *meta* isomer formed. The two nitrations (Fig. 1) occur according to the following scheme:

$$C_6H_5X \xrightarrow[\text{slow}]{NO_2^+} (NO_2)(H)C_6H_5X^+ \xrightarrow[\text{fast}]{-H^+} O_2N\text{-}C_6H_4\text{-}X$$

$$C_6H_5X \xrightarrow[\text{slow}]{NO_2^+} (H)(NO_2)C_6H_5X^+ \xrightarrow[\text{fast}]{-H^+} X\text{-}C_6H_4\text{-}NO_2$$

The field of the NO_2^+ ion induces a shift of the π-electrons toward the attacked atom r. From the point of view of quantum mechanics, such an effect can be described as an increase in the absolute magnitude of the Coulomb integral α_r of atom r ($\delta\alpha_r < 0$). The corresponding slight increase, δE_π, in the energy of the π-electrons is expressed in the LCAO MO method by

$$\begin{aligned}\delta E_\pi &= \frac{\partial E_\pi}{\partial \alpha_r}\delta\alpha_r + \frac{1}{2}\frac{\partial^2 E_\pi}{\partial \alpha_r^2}(\delta\alpha_r)^2 + \ldots = \\ &= q_r \delta\alpha_r + \frac{\pi_{rr}}{2}(\delta\alpha_r)^2 + \ldots\end{aligned} \tag{21}$$

The first term of this equation is the most important one; it means that δE_π is proportional to the π-electron charge which is localized at q_r before the start of the reaction.* Thus if we consider two reactions in which NO_2^+ attacks two different atoms of the same molecule (or atoms of the same type in molecules of different compounds), the fastest substitution should occur at atoms carrying the greatest π-electron charges q_r (minimum δE_π). In the above case such atoms are those in the *para* and *ortho* positions. The opposite holds for the nucleophilic reagents.

As has already been stated, alternant hydrocarbons are characterized by a uniform π-electron distribution over all the carbons. The first terms of expansion (21) are then identical for all the atoms, and the differences in the reactivities are due only to the second terms, $1/2\ \pi_{rr}(\delta\alpha_r)^2$. In other words, the most reactive atoms are those with the largest absolute self-polarizabilities ($\pi_{rr} < 0$). It is important to note that since δE_π is a function of the square of $\delta\alpha_r$ the reactivity of alternant hydrocarbons should be independent of whether the attacking reagent is electrophilic ($\delta\alpha_r < 0$) or neucleophilic ($\delta\alpha_r > 0$).

---

*Note that substituent X produces a non-uniform π-electron distribution in the starting C_6H_5X molecule.

As a rule, the absolute values of the self-polarizabilities of carbons in alternant hydrocarbons change hand in hand with the atomic free valencies [17]. We can thus use the free valencies calculated via the LCAO method as a measure of reactivity in heterolytic substitutions.

In free radical substitution, the attacking agent is electrically neutral, and the π-electron charges at individual atoms can no longer indicate the relative reactivities of various molecular positions.

It is probable that the unpaired electron of the radical will tend to seek out the incompletely saturated valencies in the molecule under attack to form a covalent bond. During the formation of such new bonds, the attacked carbon gradually changes from sp^2 to sp^3 hybridization. This is accompanied by partial localization of the π-electrons at r, which is equivalent to a decrease in the absolute magnitude of the resonance integrals β_{rs} of all bonds ending at r ($\delta\beta_{rs} > 0$, whatever the nature of the agent).* Effects caused by the change in the character of the π-electron localization begin to operate at a smaller radical—molecule distance than the effects of perturbance of the π-electron density.

Let us expand the change in the π-electron energy δE_π, caused by an approaching radical, into a power series in $\delta\beta_{rs}$ ($\delta\alpha_r \simeq 0$), neglecting all but the first term:

$$\delta E_\pi = \sum_s \frac{\partial E_\pi}{\partial \beta_{rs}} \delta\beta_{rs} = \sum_s 2p_{rs}\delta\beta_{rs}. \qquad (22)$$

The index s in (22) may take on all possible values. Assuming that the increments of the resonance integrals of all bonds ending

*This change in the integral β_{rs} may be explained as follows. The $2p\pi$ atomic orbital φ_r of the attacked atom tends to orient itself in the direction of maximum overlap with the valence orbital of the radical. The axis of the orbital φ_r thus ceases to be perpendicular to the plane of the molecule, and the extent of overlap of orbitals φ_r and φ_s is decreased. This, of course, leads to a reduction in the absolute magnitude of β_{rs}.

at r are the same, $\delta\beta_{rs} \simeq \delta\beta$, and bearing in mind the definition of F_r (Eq. 8), Eq. (22) can be written as

$$\delta E_\pi = \text{const} - 2F_r\delta\beta. \tag{23}$$

From this it follows immediately that, in a series of homolytic reactions of the same type, the increase in the π-electron energy δE_π will be the smallest (i.e., the activation energy will be the lowest) when the substitution occurs at that atom r which has the highest free valence F_r.

2. THE "LOCALIZATION" APPROXIMATION

The case in which this approximation is applicable is graphically illustrated by the sections of energy curves of Fig. 1 lying between points III and V. If point IV, which represents the transition complex for *meta* substitution, lies above the point IV′ (*para* substitution), then the maximum III of curve r should also be higher than the peak of curve s. In other words, the energy increment $E_{IV} - E_I$, which we shall call the atomic localization energy, is a measure of the activation energy. The electrophilic localization energy A_e^r at atom r is given by

$$A_e^r = E_e^r - E_\pi + 2\alpha. \tag{24}$$

Analogous expressions can be written for the radical (A_r^r) and nucleophilic, (A_n^r) localization energies:

$$A_r^r = E_r^r - E_\pi + \alpha, \tag{25}$$

$$A_n^r = E_n^r - E_\pi. \tag{26}$$

Here α is the energy of the bond between the localized electron and the attacked carbon C; E_π is the total energy of the π-electrons in the starting molecule, and E_e^r, E_r^r, and E_n^r are the total π-electron energies of the molecule whose conjugated system lacks atom r (the residual, or final, molecule). The differences in the values of E^r are due to unequal π-electron occupation of the upper molecular orbital of the final molecule.

Figure 1 also shows residual molecules representing the transition complexes in the substitution of singly-substituted benzene. The regions of conjugation are dotted in.

The number of mobile π-electrons in the residual molecule R depends on the nature of the attacking agent Y. If Y^- is nucleophilic, then R^- is an anion (in our example this would be the pentadienyl anion which has six conjugated π-electrons); if Y is a radical, then R is also a radical (pentadienyl radical with five π-electrons); finally, if Y^+ is electrophilic, R^+ is a cation (pentadienyl cation with four π-electrons).

The localization energies, calculated by the LCAO MO method for an alternant hydrocarbon, are identical, regardless of whether the atom is attacked by a nucleophilic, electrophilic, or radical agent (Wheland's theorem):

$$A_e^r = A_r^r = A_n^r. \tag{27}$$

Thus, for substitution in benzene we have:

$$\begin{aligned} E_\pi &= 6\alpha + 8\beta, \\ E_e^r &= 4\alpha + 2\beta(1 + \sqrt{3}), \\ E_r^r &= 5\alpha + 2\beta(1 + \sqrt{3}), \\ E_n^r &= 6\alpha + 2\beta(1 + \sqrt{3}). \end{aligned}$$

The validity of (27) is easily demonstrated by substituting these values into Eqs. (24)-(26). In agreement with theoretical predictions, the experimentally observed nucleophilic, radical, and electrophilic reactivities of alternant hydrocarbons are the same.

In computations of the localization energy by the LCAO MO method or by the free electron method, the π-electron energies of the starting molecule E_π and the residual molecule E^r are calculated separately. Dewar [38] has proposed a convenient (though only approximate) method of calculation for the alternant hydrocarbons, in which the localization energy is expressed approximately in terms of coefficients $c_{s,0}$ of the atomic orbitals of the anti-bonding molecular orbital computed for the residual molecule.

In the case of substitution reactions, Dewar's method leads to the following approximate formula for the localization energy:

$$A^r \approx 2\beta \sum_s c_{s,0}. \tag{28}$$

in which the summation involves all orbital coefficients of atoms s bonded to the attacked atom.

There is no need to calculate the energies of the molecular orbitals in the final molecule to determine the $c_{s,0}$. As has been mentioned above, these coefficients are found in the LCAO MO method from simple algebraic equations.

The free electron method may also be used to estimate reactivities within the framework of the localization approximation [37]. The calculations of the atomic localization energy then proceed from the special feature of the MO's of the π-electrons that an orbital with energy $\frac{\pi^2}{8a^2}$ is the MO corresponding to the localized electron. For example,

$$A_e^r = E_e^r - E_\pi + \frac{2\pi^2}{8a^2}.$$

Radical and nucleophilic localization energies have been determined by analogous procedures, because Wheland's theorem for alternant hydrocarbons is satisfied in the above determinations of A_e^r, A_r^r, and A_n^r by the free electron method.

As a rule, there is good agreement between the reactivities of various sites determined by usual calculations of the localization energy (via the LCAO MO method), approximate calculations of the localization energy by Dewar's method, and by the free electron method [37, 38].

3. COMPARISON BETWEEN THEORY AND EXPERIMENTAL RESULTS

Let us now compare the theoretical predictions, made on the basis of the two approximate treatments of reactivity, with the experimental data for the main classes of conjugated compounds.

Alternant hydrocarbons

In substitutions one is interested in the relative reactivities of non-equivalent sites. Calculations of F_r, π_{rr}, and A^r, for a series of alternant hydrocarbons, carried out by the simplest LCAO MO method, are shown in Table 2 [104].

Table 2

Compound	Position of atom r	F_r	π_{rr} in units of $1/\beta$	A^r in units of β	$\ln \frac{k}{k_0}$ [90]
Ethylene		0.732	0.500	2.00	—
Butadiene	1	0.838	0.626	1.644	—
	2	0.391	0.402	2.472	—
Hexatriene	1	0.861	0.685	1.526	—
	2	0.378	0.389	2.518	—
	3	0.464	0.470	2.162	—
Benzene		0.408	0.398	2.536	0
Naphthalene	1	0.452	0.443	2.299	2.69
	2	0.404	0.405	2.479	—
Anthracene	1	0.459	0.454	2.25	—
	2	0.408	0.411	2.40	—
	9	0.520	0.526	2.01	—
Diphenyl	2	0.436	0.423	2.400	—
	3	0.395	0.396	0.544	—
	4	0.412	0.411	2.447	1.26
Phenanthrene	1	0.450	—	—	2.54
	9	0.452	—	—	2.69
1,2-Benzanthracene	10	0.514	—	2.11	—
Pentacene	6	0.540	—	1.85	—
Pyrene	3	0.469	—	2.19	4.30
3,4-Benzpyrene	5	0.529	—	2.01	—

NOTE: The values of F_r and A^r are given only for the most reactive atoms.

The table indicates that position 1 of napthalene should be the most reactive, regardless of the type of attacking agent (F_1 is a maximum and A^1 minimum). This is in excellent agreement with experimental data. Note that naphthalene is one of the few aromatic hydrocarbons which can be substituted via all three mechanisms: electrophilic (nitration, sulfonation), nucleophilic (amination with sodamide), and free radical (action of diazonium salts).

The only experimental work on the other aromatic hydrocarbons with condensed rings—anthracene, phenanthrene, pyrene, 1,2-benzanthracene, 3,4-benzpyrene—involved electrophilic substitution. In virtually every case the predicted reactivities of various sites have been qualitatively confirmed by experiment.

Thus, for example, in pyrene the highest free valences and the lowest localization energies have been calculated for the equivalent atoms 3, 5, 8, and 10; and nitration, sulfonation, and methylation of pyrene do, in fact, lead to substitutions in position 3. Bromination of diphenyl gives the 4-bromo compound as predicted by the theory, and nitration results in a mixture of 2- and 4-nitrodiphenyls. Even though the F_{10} of 1,2-benzanthracene is only very slightly larger than F_9, nitration of this hydrocarbon with $HNO_3 + CH_3COOH$ produces substitution on carbon 10 exclusively.

The rates of nitration of the most reactive positions of certain aromatic hydrocarbons have been measured in relation to the rate of nitration of benzene under the same conditions. These results (Table 2) show that the logarithm of the rate constant ratio, $\ln \frac{k}{k_0}$, increases roughly in proportion to F (or A) (k_0 is the rate constant for benzene).

Radical substitutions have also been studied experimentally to confirm the theoretical predictions. Thus, the kinetics of substitution of a series of aromatic hydrocarbons with the trichloromethyl radical [39] have been investigated quite thoroughly. It was found that $\ln k_r$ was reasonably proportional to F_r in all 11 cases studied (Fig. 2).

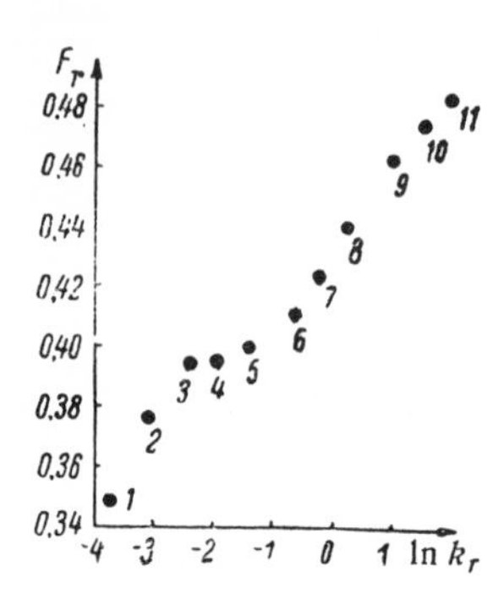

Fig. 2. The relation between the free valency F_r and the rate of substitution with radical CCl_3, (in terms of ln k_r.) 1 - benzene; 2 - phenyl; 3 - phenanthrane; 4 - naphthalene; 5 - chrysene; 6 - pyrene; 7 - stilbene; 8 - 1,2- or 5,6-dibenzanthracene; 9 - anthracene; 10 - naphthacene; 11 - 1,2-benzanthracene.

Analogous data have been obtained in substitutions with methyl [40, 41], benzyl [42], and phenyl [43, 44] radicals.

Non-alternant hydrocarbons

The theoretical interest in this class of compounds is due to the fact that the π-electron charges are distributed unequally among the carbons, while the α and β integrals of all atoms are assumed the same. In this respect the non-alternant hydrocarbons differ from conjugated systems containing heteroatoms or substituents, where additional empirical parameters must be introduced for estimation of α_X and β_{CX}.

The summary of the quantities determining the reactivities in various non-alternant hydrocarbons towards substitution reactions is given in Table 3 [45].

Relatively little information has been published so far regarding the reactivities of non-alternant hydrocarbons and their derivatives. Azulene has been studied in some detail, and the results of nitration, bromination, and acetylation confirm the theoretical prediction of electrophilic substitution into position 1 (q_1 is maximum, A_e^1 minimum) [46-48]. In contrast, the theoretical conclusion that the radical substitution should occur most readily in position 4 (F_r and A^r in agreement) conflicts with the experimental fact that a mixture of 1- and 2-benzylazulenes is formed on attack by benzyl radicals [49].

Another example of a species having an odd number of carbons in the ring, for which the theory was able to predict the relative

Table 3

Compound	Position of atom r	F_r	q_r	A_r^r in units of β	A_e^r in units of β	A_n^r in units of β
Fulvene	1	0.505	1.092	2.002	2.002	2,002
	2	0.434	1.073	2.240	2.240	2.240
	3	0.076	1,047	2,994	2.994	2.994
	6	0.974	0,623	1,612	2,230	0.994
Azulene	1	0.480	1.173	1.172	1.352	2,090
	2	0.420	1.047	1.728	1.728	1,728
	4	0.482	0.855	1.520	1,808	1.231
	5	0,429	0.986	1.655	1.655	1.655
	6	0,454	0.870	1,620	1.959	1.280
Pentalene	1	0.558	0,815	1.959	2,380	1,538
	2	0.432	1.173	2,027	2,027	2,027
Heptalene	1	0,524	1.125	1.926	1,660	2,193
	2	0.441	0,882	2.032	2.032	2.032
	3	0,494	1.119	2.026	1.715	2,357
Fulvalene	1	0.519	1.097	1.676	1,987	1,365
	2	0.499	0.976	1,844	2.236	1.452
Fluoranthene	1	0,453	0.947	—	2,466	—
	2	0.398	1,005	—	2.503	—
	3	0,740	0.959	—	2,341	—
	7	0,438	0.997	—	2.371	—
	8	0.409	1.008	—	2.435	—

reactivities of the various molecular positions is provided by the tropolonium ion [50]. The *ortho-*, *meta-* and *para-* localization energies calculated by the LCAO MO method for an electrophilic agent are 2.564, 2,686, and 2,086, respectively (in units of β). The theoretical prediction that *para*-substitution should be preferred has been confirmed by experiment.

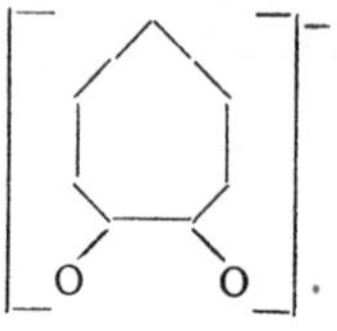

The estimates of reactivity made via the "isolated molecule"– and the "localization" approximations deviate in the case of heterolytic reactions of certain non-alternant hydrocarbons. The possible violations of the "non-intersection rule" are illustrated in Figs. 3 a–c; case 3c appears to occur during electrophilic substitution in fluoranthene. Judging by the distribution of π-electron charges, position 8 should be the most reactive (q_8 is maximum). On the other hand, the electrophilic localization energy is minimum for substitution in position 3, while position 8 is intermediate in this respect. The experiment shows that nitration in acetic anhydride gives mainly 3-nitrofluoranthene and a small amount of products substituted in positions 8, 1, and 7 [51].

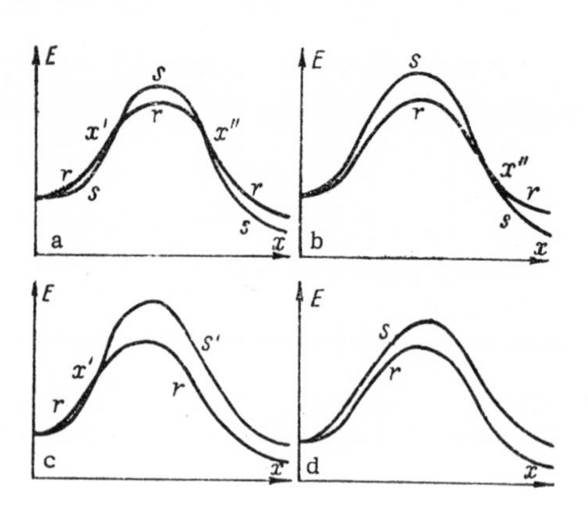

Fig. 3. Possible nature of the correlation between the energy curves of two compared reactions.

Benzene derivatives

As has already been mentioned, theoretical study of the reactivity aimed at a quantitative estimation of both the conjugation and the induction effect is considerably more difficult in the case of aromatic molecules containing substituents or heteroatoms than for conjugated hydrocarbons.

The conjugation effect is accounted for quite readily in the LCAO calculations if the presence of a substituent X capable of conjugation with the ring π-electrons is described by means of parameter β_{C-X} (which is a measure of the "strength of conjugation"). However, the numerical value of β_{C-X} cannot be obtained from purely theoretical considerations, but must be somewhat arbitrarily selected in each particular case on the basis of various considerations and empirical estimates.

Even greater difficulties in this theory are encountered if one tries to take the inductive effect into account. The simple LCAO MO method does not reflect at all the main feature of the inductive effect, which is the attenuation of atomic polarization with increasing distance from the substituent. To take this into account to some extent, the Coulomb integrals of the substituent X, of carbon C′ which carries X, and of carbon C″ which is adjacent to C′ are made slightly different than those of the remaining carbon atoms; the absolute values of the parameters characterizing these changes in the Coulomb integrals are regarded as decreasing with increasing distance from the substituent.

$$h_X > h' > h'',$$

where

$$h_X = \frac{\alpha_X - \alpha_C}{\beta}, \quad h' = \frac{\alpha_{C'} - \alpha_C}{\beta}, \quad h'' = \frac{\alpha_{C''} - \alpha_C}{\beta}.$$

Although the observed orientation of the attacking agent by the substituent can frequently be explained by introducing the parameters

h, h', h'',such an explanation is not physically acceptable. As a matter of fact, this procedure neglects the electrostatic multipole interactions with the substituent, which may in themselves explain, at least qualitatively, the most essential features of the inductive effect [52].

One of the first studies in which localization energies derived from the LCAO method were used to explain the orientation effects of substitutents was carried out by Dewar [53]. He considered the reactivities of the carbons of monosubstituted benzenes towards electrophilic agents when X = NH_2, OH, Cl, Br, F, and I. All of these are electron donors and their unshared pairs of electrons participate in the conjugation of the aromatic ring. Calculations of the π-electron energy of these molecules were carried out on the assumption that the conjugated system PhX contains eight π-electrons; the transition complex was represented by a quinonoid structure containing six π-electrons in the conjugated system of the residual molecule. The resonance integrals β_{CX} of the CX bond were assumed equal to β. The parameter h_X was given a series of values in the interval $-1 < h < 3$ so that a higher h corresponded to a higher electron affinity of the substituents.

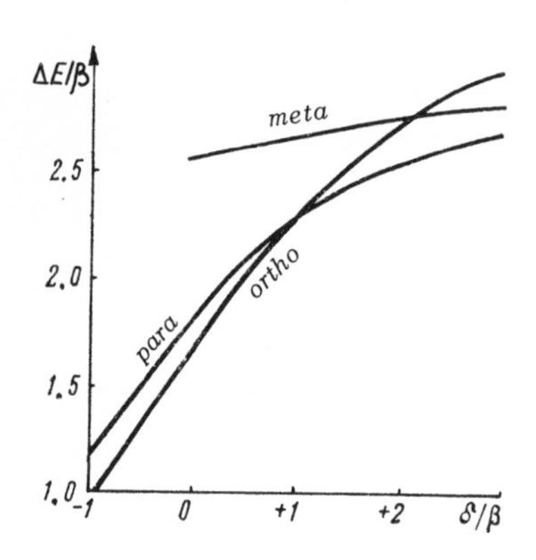

Fig. 4. Localization energies for *ortho*-, *para*-, and *meta*-substitution with F, Cl, Br, I, NH_2, etc., as a function of $h = \frac{\delta}{\beta}$ (after 53]).

To account for the changes in the electronegativities of the ring carbons caused by the inductive effect of the substituent, an additional inductive parameter $\varepsilon = \frac{1}{3}$ was introduced. It was assumed that the Coulomb integral of the carbon separated from X by n bonds was $\alpha_n = \alpha + \varepsilon^n (h\beta)$, (i.e., $h' = \varepsilon h$ and $h'' = \varepsilon^2 h$).

Localization energies calculated for various values of h in electrophilic substitution are shown Fig. 4.

The results of Dewar's calculations indicate that the characteristic features of reactions with substituted benzenes are indeed reflected by the localization approximation: substituents of this type are *ortho-* and *para-* directing towards electrophilic agents; the relative rates of substitution of singly-substituted benzenes decrease as the electronegativity of the substituent X (here characterized by increasing h) increases. Finally, according to experiment, the ratio of *ortho* and *para*-isomers should be less than unity, and the numerical value of this ratio does decrease with increasing h.

To make the theoretical results illustrated by Fig. 4 more specific, an attempt was made to find a satisfactory set of values of h for various substitutents in the series OH, NH_2, *I*, Br, Cl, F, bearing in mind that h should not be greater than 2 (i.e., $-1 < h < 2$), since otherwise the calculated molecular dipole moments deviate considerably from experimental values.

The behavior of aniline or phenol, which undergo *ortho* or *para* substitution more readily than benzene, lends itself fully to this semi-quantitative interpretation within the framework of the "localization" approximation.

As can be seen from Fig. 4, given a reasonable choice of h (e.g., $0.5 < h < 1.5$), the localization energies indicate increased ring reactivity in the *ortho* and *para* positions in aniline [A_e^0 (aniline) $< A_e^p$ (aniline) $< 2.53\,\beta = A_e$ (benzene)].

However, the MO description is not fully adequate in those cases when the inductive effect is stronger than the conjugation effect and it is difficult to introduce a second substituent. This is the situation in the halobenzenes. Predictions made on the basis of the values A_e give the correct orders of the relative reactivities in halobenzenes only if it is assumed that $h \approx 2.8$–3.2 (i.e., assuming that the electronegativities of Br, Cl, F are higher than those of N or O).

For halobenzenes, the same discrepancies with experimental data are obtained via calculation of the π-electron charges [54]. Depending on the values of h, the calculated π-electron charges in the *ortho* and *para* positions in molecules PhX vary from unity ($h \approx 3$) to considerably more than unity. The π-charge in the *meta*-position is approximately equal to unity for any h.

The isolated molecule model is thus unable to provide a satisfactory explanation for the inactivity of the chlorobenzene molecule with respect to electrophilic agents. However, the other characteristics of the reactions of PhX are given by values of q_r.

In addition to the work discussed above, various molecular-orbital calculations have been published on reactions of benzene derivatives with electron-accepting and electron-donating groups, in which the models of the substituent were further refined or different values for the parameters were introduced [55–57]. Such calculations do not fundamentally affect the explanation of the chemical reactivities in substituted benzenes as presented in this chapter.

Attempts have been made to calculate the magnitudes characterizing the reactivities of substituted benzenes by the free electron method. Thus, in [58] the reactivities of the various positions in chlorobenzene were estimated from the π-electron charge distributions, using calculations [59] in which the presence of the Cl atom in the conjugated system was accounted for by introducing in its place a rectangular depression at the bottom of the potential well. The magnitude of this depression was taken as equal to the difference in electronegativity between chlorine and carbon.

Conjugated systems containing heteroatoms

Quantum-chemical studies of the reactivities of such systems have been limited almost exclusively to nitrogen heterocycles containing five or six atoms in the ring. Most of the work dealt with pyridine [60-64]. The π-charges on various atoms in this compound have been computed by the usual LCAO MO method of several values of the integral β_{CN} and of the prameter $h = \frac{\alpha_N - \alpha_C}{\beta}$, which characterizes the increased electronegativity of the N atom ($h > 0$). It was found that, over a wide range of h and β_{CN}, the π-charges on atoms 2 and 4 (the nitrogen being atom 1) were less than unity, while the charge on atom 3 was slightly less than or approximately equal to unity, so that the pyridine ring would be less active than benzene.

The values of parameters of reactivity of pyridine calculated for $\beta_{CN} = \beta$ and two values of h, are given in Table 4. One of these states (h = 0.5) fits the free pyridine molecule, while the other (h = 2.0) fits pyridine in the form of a conjugated acid in an acid medium [64].

Table 4

h	Position of atom r	A_e^r in units of β	A_r^r in units of β	A_n^r in units of β	q_r	F_r
0.5	2	2.6718	2.5124	2.3529	0.923	0.399
	3	2.5381	2.5381	2.5381	1.04	0.398
	4	2.7011	2.5374	2.3737	0.950	0.402
2.0	2	2.7087	2.2838	1.8588	0.759	0.515
	3	2.5604	2.5604	2.5604	1.012	0.488
	4	3.0711	2.5418	2.0125	0.835	0.438

According to theoretical calculations, bromination, nitration, and sulfonation of pyridine should give 3-substituted products, while amination or methylamination should occur at the 2-position.

Radical substitution, which occurs when pyridine is attacked by phenyl radicals, leads to the formation of α-phenylpyridine, with a small proportion of the β- and γ-isomers [65]. This result disagrees with theoretical predictions based on the calculated values of F_r (Table 4).

The reactivities of quinoline have been calculated for the same values of the parameters as those used with pyridine, that is, $\alpha_N = \alpha + 0.5\beta$ for the free base and $\alpha_N = \alpha + 2\beta$ for the protonated molecule [69]. They are shown in Table 5, together with the corresponding experimental data. There is a lack of agreement between the predictions based on π-electron charges or the localization energy and the actual reactivities with respect to electrophilic agents in moderately acid media.

Table 5

Substitution	Calculations	Observed results
Electrophilic	1) π-electron charges show that the reactivity should be highest in position 8 and then in 3 ($h = 2$) 2) Localization energy predicts first an attack on position 8, then on 5 ($h = 2$)	In strongly acid media the nitration proceeds in positions 8 and 5; sulfonation chiefly in 8, less so in 5; in less acidic media the reactivities decrease in the order $3 > 6, 8$
Nucleophilic	1) Predictions from values of q_r indicate first attack in position 2, then in 4 ($h = 0.5$) 2) Localization energy predicts an attack in position 4, then on 2 ($h = 0.5$)	Amination and hydroxylation are directed mainly to position 2, slightly to 4.
Radical	1) Indices F_r indicate that the reactivities of positions should decrease in the order: $4 > 2 > 5$ 2) Localization energy shows that: $-4 \approx 5 \approx 8$	Reaction of phenyl radical with benzoyl peroxide leads to 8-substitution and smaller quantities of other isomers ($8 > 9 > 3 > 5 > 6$; $7 > 2$)

To explain the formation of 3-substituted quinoline, the intermediate formation of a dihydroquinoline system is postulated; this is then supposed to be attacked by the electrophilic agent at the point of maximum π-electron density [66, 67]. To find out what intermediate system can in fact be formed, quantum-mechanial calculations of π-electron densities in the assumed intermediate 1,4- and 2,3-dihydroquinolines have been made [67], and show that a maximum π-charge in position 3 can occur only in the case of 1,4-dihydroquinoline.

Theoretical work was done on the reactivities of the carbons in other N-containing heterocycles: pyrimidine, pyrazine [64], glyoxaline, pyrazole, pyrrole [68-71], and some others. In pyrimidine, at $h = 0.5$ and $h = 2.0$, the reactivities for electrophilic substitution decrease in the order of $5 > 4 > 2$ (A_e^r and q_r are in agreement), and for nucleophilic substitution in the order of $2 > 4 > 5$ (A_n^r and q_r are in agreement). These predictions are indirectly confirmed by experimental observations on pyrimidines containing activating groups. Thus, nitration, halogenation [72, 73], and nitrosation [74] occur in position 5; amination of 4-methylpyrimidine gives the 2,4-diamino derivative.

In five-membered heterocycles such as pyrrole, glyoxaline, and pyrazole, at two values of h [$h = 1$ (neutral molecule) and $h = -1$ (anion)], the directions of nucleophilic and radical substitutions predicted from values of q_r and A_n^r, F_r and A_r^r are in agreement. An exception is the radical reactivity of the pyrrole anion [70].

For the neutral pyrrole molecule ($h = 1$) the predictions based on consideration of A_e^r and q_r, disagree with each other. Calculations of the π-electron charges indicate that electrophilic substitution in a neutral pyrrole molecule should occur in the α-position. The minimum of A_e^r corresponds to localization in the β-position. In actual fact, in acid media electrophilic substitution is directed to the α-position. We can say that the energy curves of the reaction intersect, as shown in Fig. 3b. Analogous violations of the "non-intersection" rule occur in the electrophilic substitution into glyoxaline and pyrazole anions. By contrast, in the cases of neutral glyoxaline and pyrazole molecules (h = 1), both approximations of the theory agree with each other and with the experiment.

The satisfactory agreement obtained between calculated reactivities for the anions of the five-membered heterocyclics indicates that in conjugated systems in which the nitrogen carries a formal negative charge its negativity can be assumed smaller than that of carbon [75, 77].

The theoretical studies on the reactivities of non-alternant, N-containing heterocycles deserve some attention [76]. The results obtained by the two approximate treatments show that placement of a nitrogen atom in any position of azulene (in place of a CH group) should not affect the direction of ionic substitution. These predictions require experimental confirmation.

ADDITION REACTIONS

1. THE "ISOLATED MOLECULE" APPROXIMATION

A typical example of 1,4- addition (*para*-addition) is the simultaneous attack of maleic anhydride on two atoms, a and b, of a benzene ring, these atoms being *para* to each other. In the intermediate stage of the reaction, when the double bond between atoms 2 and 3 in the anhydride is being opened, the latter may be regarded as a pair of free radicals.* Therefore, as in the case of the homolytic substitution reactions, we can use the charge of energy caused by partial localization of π-electrons on the attacked carbons as a criterion of reactivity. Decreased delocalization is reflected by a small reduction of the values of integrals β between the attacked atoms a and b and their neighbors.

*Various mechanisms have been proposed for the diene synthesis, among them the heterolytic, homolytic, and cryptoradical (further details can be found in [12]).

Then

$$\delta E_\pi = \frac{\partial E_\pi}{\partial \beta_{ac}} \delta\beta_{ac} + \frac{\partial E_\pi}{\partial \beta_{ad}} \delta\beta_{ad} + \frac{\partial E_\pi}{\partial \beta_{be}} \delta\beta_{be} + \frac{\partial E_\pi}{\partial \beta_{bf}} \delta\beta_{bf}. \tag{29}$$

Assuming that $\delta\beta_{ac} = \delta\beta_{ad} = \delta\beta_{be} = \delta\beta_{bf} = \delta\beta$, we obtain

$$\delta E_\pi = \text{const} - 2(F_a + F_b)\,\delta\beta. \tag{30}$$

The following addition rule for Diels-Alder reactions may be established on the basis of (30): in additions to the *para*-positions of a series of homologs, the rate is highest in the compound in which the sum $(F_1 + F_4)$ has the largest value.

In the "isolated molecule" approximation, the reactivity of conjugated hydrocarbons with respect to *ortho*-addition can be characterized by two methods. In the first case, one uses the indices of bonds *ac* and *bd* which are adjacent to the attacked bond *ab*; in the second case, one employs one of the indices of the attacked bond itself.

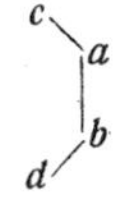

In the first approach we can again assume that the effect of the attacking agent consists in reducing the value of the resonance integrals of bonds *bd* and *ac*. Limiting ourselves to the first terms of the expansion, we find

$$\delta E_\pi = 2(p_{ac} + p_{bd})\,\delta\beta. \tag{31}$$

For example, the rate of reaction A + OsO_4 (where A is an aromatic hydrocarbon) should be highest when the sum of the orders of bonds adjoining the bond *ab* is minimum (i.e., when the sum of the free valences is maximum).

In the second case we start with the assumption that the π-electron charge between atoms a and b increases during *ortho*-additions. This increase in the double-bond character of ab is reflected in an increase in the absolute value of the resonance integral β_{ab}. The corresponding energy increment is

$$\delta E_{\pi} = \frac{\partial E_{\pi}}{\partial \beta_{ab}} \delta \beta_{ab}. \tag{32}$$

Expression (32) shows that, for example, the OsO_4 molecule will selectively attack the highest-order bonds. On comparison with the experimental data, both indices of reactivity, that is, $(F_a + F_b)$ and p_{ab}, give similar results; in practice, the order of the attacked bond is the more convenient criterion of reactivity in *ortho*-additions.

2. THE "LOCALIZATION" APPROXIMATION

It is assumed that the Diels-Alder reaction proceeds via a transition complex having the same structure as the final product. Thus, in the addition of maleic anhydride to anthracene which is typical of these reactions:

$$\text{(anthracene)} + \begin{matrix} CH-CO \\ \| \\ CH-CO \end{matrix} \!\!> O = \text{(adduct)} \begin{matrix} CH-CO \\ | \\ CH-CO \end{matrix} \!\!> O$$

The conjugated system of the final molecule in this reaction consists of two isolated benzene rings. The localization involves two π-electrons: these localize on atoms to which the dienophile adds, and form σ-bonds; the other π-electrons of the aromatic molecule become delocalized in the conjugated system still remaining after localization.

This concept of the process defines the *para*-localization energy, B_p^{mn}, which is used as a reactivity index in Diels-Alder reactions [77]:

$$B_p^{mn} = E_p^{mn} - E_\pi + 2\alpha, \tag{33}$$

where, as before, E_π is the total energy of the π-electrons in the starting molecule, and E_p^{mn} is the π-electron energy of the conjugated systems obtained after localization of two π-electrons at atoms m and n which are *para* to each other.

The reactivity of aromatic hydrocarbons with respect to such agents as osmic acid or ozone is characterized by a localization energy (or *ortho*-localization) B_0^{mn}, equal to the energy necessary for the conversion of the attacked bond into an isolated double C=C bond [78, 79]:

$$B_0^{mn} = E_0^{mn} - E(=) - E_\pi, \tag{34}$$

where $E(=)$ is the energy of the isolated double bond and E_0^{mn} is the π-electron energy of the final molecule obtained from the starting molecule by isolation of atoms m and n which form the double bond (the resonance integrals of bonds adjacent to the attacked one become zero). In the 9, 10 addition of OsO_4 to phenanthrene:

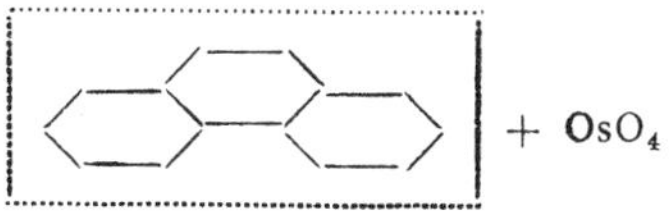

$+ OsO_4$

the region of free circulation of the π-electrons in the transition complex is the same as that in the molecule of diphenyl:

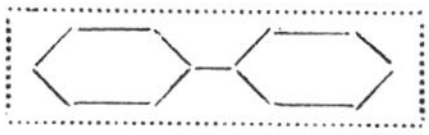

The localization energy of bond 9,10 can be found by subtracting the π-electron energy of the ground state of the phenanthrene molecule from the sum of the π-electron energies of diphenyl and ethylene.

In alternant hydrocarbons, the method of the anti-bonding MO allows the estimation of energies B_p and B_0 without having to solve equations for the π-electron energies [38]. As in the case of the substitution reactions, the localization energies for additions are expressed in terms of coefficients of the AO's.

The energies of *para-* and *ortho-* localization can be expressed in terms of the conjugation energies of the starting (R^*), "final" (R_p^{*mn}) or R_0^{*st} molecules:

$$B_p^{mn} = R^* - R_p^{*mn} - 2\beta, \tag{35}$$

$$B_0^{st} = R^* - R_0^{*st}. \tag{36}$$

Equations (35) and (36) are obtained directly from Eqs. (33) and (34) for the localization energy and from the equations for the conjugation energy. Let us recall that in the LCAO MO method the conjugation energy of a molecule with $2n$ π-electrons is equal to the difference between the π-electron energy of the conjugated bond system and the energy of a set of isolated π-bonds:

$$R^* = E_\pi - nE(=). \tag{37}$$

The conjugation energy can be estimated by various approximate methods, without becoming involved in laborious calculations [79]. Often, an empirical method is used in which the conjugation energy of a complex hydrocarbon RS is determined from the conjugation energies of two simpler systems R and S [79, 80].

3. COMPARISON OF THEORY WITH EXPERIMENT

The reactivities of carbons in *para*-positions with respect to additions of dienophilic reagents have been studied in polyacenes. As can be seen from Table 6 [104], there is full agreement between the reactivities calculated by the standard LCAO MO method from the sums of the free valences of the attacked atoms and those from the *para*-localization energies. It also follows that, as the number of linearly fused benzene rings is increased, the relative reactivities of the *para*-positions should also increase (from benzene to pentacene). In the case of an angular fusion of rings, on the other hand, the reactivity for the addition of dienophiles should decrease by comparison to the corresponding linear isomer.

Table 6

Compound	Attacked *para*-atoms	$\sum_r F_r$	*Para*-localization energy
Benzene	1— 4	0.796	4.0
Naphthalene	1— 4	0.904	3.68
Anthracene	1— 4	0.918	3.68
	9—10	1.040	3.31
Naphthacene	5—10	1.060	3.25
Pentacene	5—14	1.062	3.23
	6—13	1.080	3.18
1,2-Benzanthracene	9—10	1.016	3.41
1,2,5,6-Dibenzanthracene	9—10	0.996	3.51
Phenanthrene	1— 4	0.892	3.77

The theoretical predictions are in good agreement with experimental data. Benzene, naphthalene, and phenanthrene, all characterized by low values of $\sum_r F_r$, do not react at all with maleic anhydride [81, 82]. The latter adds with increasing ease in the series from anthracene to naphthacene and pentacene (central *meso* positions). If the observed rate of addition to the *para*-atoms 9,10 in anthracene is taken as unity ($k = 1$), then $k = 30$ for the 5,12 positions in naphthacene, and 1600 for the 6,13 positions in pentacene [83]. The additions of maleic anhydride to non-linearly fused ring systems, e.g., to 1,2-benzanthracene and particularly to 1,2-5,6-dibenzanthracene, are slower than to anthracene.

Values of p_{rs} and B_0^{rs} for various bonds of alternant hydrocarbons, calculated by the LCAO MO method, are listed in Table 7 [17].

The experimental data on the positions at which OsO_4 or O_3 add to condensed aromatic molecules largely confirm the theoretical calculations.

Table 7

Compound	Attacked bond	p_{rs}	*Ortho*-localization energy
Benzene	1— 2	0.667	3.53
Naphthalene	1— 2	0.725	3.26
	2— 3	0.603	3.72
Anthracene	1— 2	0.738	3.22
	2— 3	0.586	3.80
Naphthacene	1— 2	0.741	3.21
	2— 3	0.581	3.88
Chrysene	1— 2	0.754	3.32
3,4-Benzophenanthrene	1— 2	0.762	3.10
Phenanthrene	9—10	0.775	3.07
Pyrene	4— 5	0.777	3.06
1,2-5,6-Dibenzanthracene	3— 4	0.778	3.05
1,2-Benzanthracene	5— 6	0.783	3.03
Pentacene	6— 7	0.790	3.01

In accordance with the theory, the rate of the addition of OsO_4 to the 1,2-bond in naphthalene is higher than that to benzene (in fact, benzene does not react with OsO_4); the rate of addition to the 1,2-bond in anthracene is greater than in naphthalene, and addition to the 9,10 bond in phenanthrene is much faster than that to naphthalene or anthracene [84, 85].

Assuming that the relative activation energies $\Delta E_2 - \Delta E_1$ in addition reactions are functions of the corresponding *ortho*-localization energies, Brown [80] has calculated the relative rates of these reactions by the formula

$$\frac{k_1}{k_2} = \exp\frac{\Delta E_2 - \Delta E_1}{RT} \quad (RT \sim 0.6 \text{ kcal}).$$

The calculated rates of addition to a number of aromatic molecules were compared with those to the most reactive bond of 1,2-benzanthracene which was taken as unity. The results were in good agreement with experiment (Table 8):

Table 8

Compound	Theor. k_1/k_2 [80]	Exptl. k_1/k_2 [85]
1,2-Benzanthracene	(1)	(1)
Phenanthrene	0.2	0.1
3,4-Benzopyrene	2.0	2.0
1,2-5,6-Dibenzanthracene	1.0	1.3
5,6-Benzochrysene	0.04	very small

The experimental data on the ozonization reactions cannot always be interpreted with the aid of indices p_{ik} and E_0. In agreement with theoretical predictions, a number of aromatic hydrocarbons (naphthalene, pyrene, and others) add ozone in positions of maximum p_{ik} and minimum E_0. However, ozonizations are not simple *ortho*-additions, as shown by the fact that one of the main oxidation products of anthracene is 9,10-anthraquinone. For this reason, many two-stage mechanisms have been proposed to reconcile ozonization with the MO treatment. One of these mechanisms postulates that a π-complex is first formed between the aromatic ring and the ozone molecule [86]:

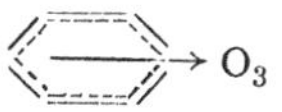

$\rightarrow O_3$

The reaction can now be directed either into positions 1,2 (with cleavage of the π-bond between the carbons, or to 1,4 (quinone), depending on which direction ensures the minimum reduction in the conjugation energy. Thus, this theory postulates that in each specific case, the reactivity index should be the smaller of the two values E_0 or E_n.

Other authors [87-89] suggest that the attacking ozone acts as an electrophilic agent to form a complex, and that the oxidation products result from a rearrangement of that primary complex. According to Wibaut [88], the attacking electrophile is the central positive oxygen of the ozone molecule. In contrast, Meinwald [87] assumes that the attacking ozone atom is a terminal one.

One difficulty in determining the true mechanisms of ozonizations is due to the fact that it is influenced by the nature of the solvent.

A BRIEF ACCOUNT OF OTHER THEORETICAL RESULTS

1. HAMMOND'S RULE

Up to now we have compared the reactivities of various classes of organic compounds in reactions of the same type, always involving the same attacking agent. Analysis of the experimental data shows that in such cases there usually holds a linear relationship of the form

$$\log \frac{k_2}{k_1} = \frac{\Delta E_1 - \Delta E_2}{RT}. \tag{38}$$

However, the slopes of the plots of $\log \frac{k_2}{k_1}$ against $\frac{1}{T}$ vary according to the nature of the agent attacking the aromatic hydrocarbon. This fact contradicts Wheland's transition complex model because Wheland assumed that the reactivities of alternant hydrocarbons in any given type of reaction are independent of the attacking agent.

The slopes of these plots and the corresponding expression for ΔE_π^- [90] can be used to determine the values of the resonance integral β. As a rule, the β found in this way tend to be lower than the normal value (20 kcal/mole), and this difference increases with the reactivity of the attacking agent. In other words, as the chemical activity of the attacking agent increases, the transition complex hybridization of the attacked carbon can be described better by an sp^2- than an sp^3 state (i.e., it differs more and more from the state postulated in Wheland's transition complex).

The results reported in [90] and certain other experimental data show that the following rule, formulated by Hammond [91], is widely applicable to aromatic substitution: *in fast reaction the electronic structure of the transition complex resembles the structure of the starting reactants (that is, the structure prior to reaction); in slow reactions, the structure of the complex resembles that of Wheland's model.*

This rule readily explains the fact that when quinoline is substituted by "strong" nucleophiles such as NH_2^- or OH^-, the direction of the reaction appears to be related solely to the π-electron charges (Table 5). It also allows us to interpret correctly the experimental data on variation of the *ortho/para* ratio as a function of the nature of the substituent in the benzene ring, the attacking agent [92], and so on.

2. COMPARISON OF VARIOUS REACTIVITY INDICES

There has been no lack of attempts at verifying the above theories and their modifications. These postulated that reactivity correlates not only with the properties discussed above but also with other theoretical quantities characterizing the electronic structure; or they attempted to unite the "isolated molecule" and the "localization" approximations into a single theory of chemical reactivity. We shall now outline, in very general terms, the various directions in which the theory has developed.

Among the theories which base the reactivity on the electron structure of the starting unperturbed molecule, let us indicate those in which attempts have been made to correlate the reactivity of conjugated hydrocarbons with the charge distribution on the carbons created by the π-electrons of individual molecular orbitals. Thus according to Fukui and Yonesava [93, 94] the quantity responsible for the reactivity of a carbon atom in electrophilic substitutions is the density of π-electrons occupying the "boundary" (highest occupied) energy level.

Indeed, the most reactive atoms of alternant hydrocarbons (terminal atoms of butadiene, α-atoms in naphthalene, *meso*-atoms in anthracene, etc.) also exhibit the highest π-electron densities at that level (as calculated by the standard LCAO MO and free electron methods) while the overall π-electron charges are the same on all the atoms of the conjugated systems.

Fujui and Yonesava cite the relative ease with which this index of reactivity may be calculated.

In addition to the usual "localization" approximation, in which the activation energy of the reaction is characterized by the localization energy calculated by Wheland's method, several other models of the transition complex have been proposed as follows:

1. A model assuming delocalization of the π-electrons in aromatic substitutions which takes into account hyperconjugation between the so-called "psuedo π-bond -C-HY" and the π-electron system of the remainder of the molecule [95, 96].

2. A model postulating the transfer of an electron [97-99].

In the first of these models the index of reactivity is the "hyperconjugation energy," $E_{\text{hyperconjug}}$, defined as:

$$E_{\text{hyperconjug}} = E_\pi(\text{benzene}) - E_\pi(\text{C}_6\text{H}_5\text{–HY}).$$

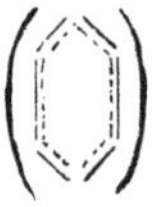

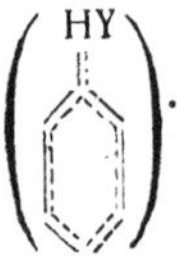

The introduction of the concept of the "psuedo π-bond" in the transition state is an attempt to reflect the ability of the π-electrons to shift from ring to the reagent during the reaction.

Muller, Pickett and Mulliken [95] have shown that the introduction of hyperconjugation produces additional stabilization of the transition complex model in the isotopic-exchange of the hydrogen present in the CH bonds of benzene. Hyperconjugation appears to have a smaller effect on the activation energy of other substitution reactions [96].

In a transition-complex model which assumes hyperconjugation, the relative reactivities can be estimated by expanding into a power series the changes in the π-electron energy in the vicinity of the point which represents the assumed model on the energy curve for the reaction (just as in the "isolated molecule" approximation for the initial point of the energy curve). The index of reactivity for substitution reactions then becomes the magnitude of "hyperdelocalizability" [100]. Large values of this quantity indicate high activity of the carbon under consideration. Until recently, the use of "hyperdelocalizability" as an index of reactivity in large molecules was difficult because it required the solution of an algebraic equation of high order and the determination of the coefficients of the LCAO MO method. The situation was considerably simplified after the appearance of Mestechkin's paper [101], which proposed a new and efficient method for computing this quantity.

In concluding this brief survey of delocalization models, let us note that the concepts of conjugation and hyperconjugation have been subjected to criticism by certain authors [102]. In particular, the applicability of the concept of hyperconjugation to electronic structure of reacting molecules has been questioned. It was shown that many physical and thermochemical properties of organic molecules can be explained without hyperconjugation. All that needs to be done in those cases is to account properly for the effect of hybridization of the atomic level on bond lengths and energies.

A more detailed discussion of this problem would be outside the scope of this review. We think that it would be desirable to devise a model in which the effects of conjugation, hyperconjugation, and hybridization would complement each other in describing the behavior of molecules during reactions.

The other working hypothesis is that of electron transfer. Thus, if an electrophilic substitution is to occur, the lowest free energy level of the attacking agent must lie below the highest occupied level of the attacked hydrocarbon; the electron of the hydrocarbon is then readily transferred to the attacking agent. For benzene and methane, the highest occupied levels lie at -9.24 and -13.0 eV, respectively. In electrophilic agents NO_2^+, I^+, Br^+, Cl^+, and H^+, the lowest free levels are -11.0, -10.4, -11.8, -13.0, and -13.6 eV. In Ag^+ and I_2^{++}, which do not react with benzene, these levels are at -7.6 and -7.0 eV. Thus, the above hypothesis explains why saturated hydrocarbons like methane are inert to most of the electrophilic agents which readily react with benzene.

In the electron-transfer hypothesis, the model of the transition complex for an aromatic substitution is a system consisting of a hydrocarbon cation (e.g., $C_6H_6^+$) and a neutral particle (e.g., a bromine atom). The relative activity of the reaction centers is determined from the free valences of atoms of the cation left as a result of the electron transfer.

A more rigorous treatment of reactivity via the charge-transfer model has been suggested by Brown [97]. In this case, the transition state is stabilized by a partial transfer of the π-charge from the aromatic molecule to the electrophilic agent, without any modification of the π-electron system. The calculation of the resulting transition complex models by the perturbation method leads to a new reactivity index Z, which includes two empirical parameters Y_e and $g_e^{\neq}$. Using the nitration of aromatic hydrocarbons as an example, Brown showed that there is a certain correlation between Z and the rate constants of these reactions.

However, the mean deviation of the calculated points from the straight line (6%) is large, considering that the expression for Z includes two *empirical* parameters.

As far as the reactivities of alternant hydrocarbons are concerned, all the theoretical approaches are in close agreement. This has been noted by several authors [38, 102, etc.], but the mutual relationship of the proposed methods for calculation of reactivity has been analyzed in detail only in relatively recent times. On the basis of the LCAO MO theory, Baba [103] has shown that, when a certain condition for the secular determinants of the final molecules is satisfied, the results obtained via the "isolated molecule" and the "localization" approximations should coincide. Baba's calculations show that the above condition is satisfied, in fact, in many even alternant hydrocarbons. In the same paper, Baba proposed a new generalized treatment of reactivity of molecules containing conjugated bonds; in this treatment, the changes in the π-electron energy can be compared at any intermediate stage of a series of reactions of the same type. The corresponding reactivity index P_r, the partial polarization energy, is regarded as a function of the variable π-electron charge. In the starting state, the charge on the attacked atom r is $q_r = 1$ and $p_r(q_r) = 0$; in the transition complex formed during the electrophilic substitution, $\lim_{q_r \to 2} p_r(q_r) = A_e$; in nucleophilic substitution $\lim_{q_r \to 0} p_r(q_r) = 0$.

A more general discussion of the reasons why the results of various methods of estimating the reactivity coincide is given by Fukui et al. [104]. In the case of alternant hydrocarbons and for heteroalternant molecules, these authors have given a mathematical formula for the condition at which reactivities calculated on the basis of self-polarizability, free valence, density of the boundary electrons, hyperdelocalizability, and Wheland's or Dewar's atomic localization energies are equivalent.

3. APPLICATION OF THE THEORY TO DISSOCIATION AND POLYMERIZATION REACTIONS

The MO theory of chemical reactivity can also be applied to reactions other than those considered in the previous sections.

For example, the literature contain many papers on attempts to apply the theory to dissociation of conjugated molecules into ions or free radicals. In contrast to the substitutions and additions, the transition complex can, in this case, be represented by a model involving a greater degree of delocalization of the π-electrons than that in the starting conjugated molecule.

The ionization of allyl chloride in an aqueous medium is a typical example:

$$CH_2{=}CH{-}CH_2Cl \rightarrow (CH_2{\cdots}CH{\cdots}CH_2)^+ + Cl^- \,(aq).$$

This ionization to the unstable carbonium ion $(CH_2{\cdots}CH{\cdots}CH_2)^+$ appears to be the slowest step of the hydrolysis

$$CH_2{=}CH{-}CH_2Cl + H_2O \rightarrow CH_2{=}CH{-}CH_2OH + HCl.$$

The MO method allows us to estimate the changes in the rate constant of the ionization of allyl chloride in comparison with the ionization of the analogous (but saturated) molecule of *n*-propyl chloride. All that is required is that we make the usual assumption that the contributions of σ-electron energies to the two activation processes are the same, $\Delta E_\sigma^q = 0$. The changes in the π-electron activation energy, ΔE_π^a, are taken as equal to the differences between the π-electron energies of the $(CH_2{\cdots}CH{\cdots}CH_2)^+$ ion and of the starting nonionized molecule $CH_2{=}CH{-}CH_2Cl$ [12]

$$\Delta E_\pi^a = E_\pi (CH_2{\cdots}CH{\cdots}CH_2)^+ - E_\pi (CH_2{=}CH{-}CH_2Cl) =$$
$$= (2\alpha + 2.828\beta) - (2\alpha + 2\beta) = 0.828\beta = 16.5 \ \frac{\text{kcal}}{\text{mole}}$$

$$\left(\beta = 20 \ \frac{\text{kcal}}{\text{mole}}\right).$$

The 16.5 kcal/mole difference between the activation energies for ionization of allyl chloride and *n*-propyl chloride should mean that the rate of the former is higher by a factor of the order of 10^{12}. The rate constants have not been experimentally measured, but an analogous estimate proved correct in the case of the free radical dissociations of allyl iodide and ethyl iodide, for which the experimental dissociation energies proved to be 36 and 51 kcal/mole, respectively [105], in good agreement with the theoretical estimate [106].

Another example of the use of the "delocalization" approximation in LCAO MO calculations is the ionization of triphenylmethyl chloride and its derivatives in liquid SO_2 [107]. The agreement between theory and experiment is quite satisfactory. In particular, it has been shown that when a phenyl group is introduced into the *meta*-position of the benzene ring, the ionization of triphenylmethyl chloride should be suppressed, since the π-electron energy of the starting compound is stabilized to a greater extent than in the cation.

A simplified treatment of free radical polymerization and copolymerization has also been developed [108] within the framework of the "delocalization" approximation. In that case, it is assumed that the rate constant for chain propagation, the rate constant for the competing branching reaction, and also the structure of the resulting polymer depend mainly on the energy of the additional (as compared with the initial system) conjugation of the π-electrons in the transition complex. Thus if during the polymerization the rth atom of the monomer interacts with the sth atom of the radical, then the rate of the reaction increases with the π-electron conjugation energy $\Delta E_{r,s}$, caused by this interaction. A perturbation method was used to estimate the conjugation energy in this transition complex, assuming the additional conjugation $\Delta E_{r,s}$ between the monomer and the radical to be a small perturbation. The authors of [109] calculated values of $\Delta E_{r,s}$ for various positions of monomers and radicals of styrene, butadiene, vinyl chloride, and acrylonitrile.

In [110], the above treatment is used to determine the energetically most favorable structures of polymers and copolymers of butadiene, chloroprene, and acrolein.

A transition complex model involving the delocalization of electrons has recently been proposed by Ya. K. Syrkin to explain the mechanisms of acid-base catalysis, molecular rearrangements, hydrolysis, esterification, isotopic exchange, and other organic reactions of molecules which contain no conjugated bonds but do include isolated double bonds or unshared pairs of electrons [111].

According to Syrkin, the most favorable reaction path is via formation and decomposition of a cyclic transition complex, in which the old bonds are breaking up but no new bonds are as yet formed. The proof of an appreciable reduction of the activation energy for 7-, 6-, and 5-membered rings containing 6 electrons is based on a large number of examples. Unfortunately, quantitative calculations by the MO method cannot as yet be adapted to such rings when they contain σ-bonds.

REFERENCES

1. S. Glasstone, K. Laidler and H. Eyring. Theory of Absolute Reaction Rates [in Russian], Foreign Literature Press, 1943.
2. M. G. Veselov. Elementarnaya Kvantovaya Teoriya Atomov i Molekul [Elementary Quantum Theory of Atoms and Molecules], Moscow, Fizmatgiz, 1962.
3. Ya. K. Syrkin. Dokl. Akad. Nauk SSSR, 1955, 105, 1018.
4. Kh. S. Bagdasar'yan. Zh. Fiz. Khimi, 1949, 23, 1375.
5. R. Daudel and O. Chalvet. Calcul. fonct. onde molecul. Paris CNRS, 1958, 389.
6. G. Wheland. JACS, 1942, 64, 900.
7. H. Seel. Angew. Chem., 1948, 60, 300.
8. R. D. Brown. Quart. Rev., 1952, 6, 63.

9. C. A. Coulson. J. Chem. Soc., 1955, 2078.
10. O. A. Reutov. Teoreticheskie Problemy Organicheskoi Khimii [Theoretical Problems of Organic Chemistry], Press of Moscow University, 1955.
11. Sostoyanie Teorii Khimicheskogo Stroeniya v Organicheskoi Khimii [The State of the Theory of Chemical Structure in Organic Chemistry], Press of Acad. Sci. USSR, 1954.
12. John Roberts. Notes on Molecular Orbitals Calculations, N. Y., 1961.
13. C. A. Coulson and H. C. Longuet-Higgins. Revue Scientifique, Paris, 1947, 85, 939.
14. C. A. Coulson. Research, 1951, 4, 307.
15. R. Daudel, R. Lefebvre and C. Moser. Quantum Chemistry Methods and Applications, N. Y., 1959.
16. W. A. Bingel. Naturforsch., 1954, 9A, 5, 436.
17. B. Pullman and A. Pullman. Les theories electroniques de la chimie organique, Paris, 1952.
18. C. A. Coulson. Valence, Oxford, 1961.
19. H. Hartmann. Theorie der chemischen Bindung auf quantentheoretischer Grundlage, 1954.
20. U. Kozman. Introduction to Quantum Theory [in Russian], Moscow, Foreign Literature Press, 1960.
21. C. A. Coulson. Quart. Rev. (Chem. Soc.), 1947, 1, 144.
22. C. A. Coulson. Determination of Organic Structures by Physical Methods, N. Y., 1955.
23. M. V. Vol'kenshteyn. Stroenie i Fizicheskie Svoistva Molekul [Structure and Physical Properties of Molecules], Press of Acad. Sci. USSR, 1955.
24. H. Bayliss. Quart. Rev. (Chem. Soc.), 1952, 6, 319.
25. G. Wheland. Resonance in Organic Chemistry, N. Y., 1955.
26. G. Wheland and L. Pauling. JACS, 1935, 57, 2091.
27. C. A. Coulson. Proc. Roy. Soc., 1939, A169, 413.
28. C. A. Coulson. Proc. Roy. Soc., 1938, A164, 393.
29. M. J. Dewar and R. Pettit. J. Chem. Soc., 1954, 1617.
30. C. A. Coulson. Disc. Far. Soc., 1947, 2, 9.
31. W. E. Moffitt. J. Chem. Phys., 1948, 45, 243.
32. W. E. Moffitt. Trans. Far. Soc., 1949, 45, 373.
33. K. Ruedenberg and C. Sherr. J. Chem. Phys., 1953, 21, 1565.
34. C. A. Couslon and H. S. Longuet-Higgins. Proc. Roy. Soc., 1948, A195, 188.
35. C. A. Coulson and G. S. Rushbrooke. Proc. Gamb. Phil. Soc., 1940, 36, 193.
36. C. A. Coulson and H. C. Longuet-Higgins. Proc. Roy. Soc., 1947, A191, 16.
37. M. N. Adamov and I. F. Tupitsyn. Vestnik LGU, 1962, No. 16, 47.

38. M. J. Dewar. JACS, 1952, 74, 3355.
39. E. C. Kooyman and E. Farehnorst. Trans. Far. Soc., 1953, 49, 58.
40. M. Levy and M. Szwarg. JACS, 1954, 76, 1949.
41. M. Levy, M. S. M. Newman and M. Szwarg. JACS, 1955, 77, 4225.
42. J. R. Dunn, W. A. Waters and I. M. Roitt. J. Chem. Soc., 1954, 580.
43. D. H. Hey and G. H. Williams. Disc. Far. Soc., 1953, 14, 216.
44. D. H. Hey and G. H. Williams. J. Chem. Phys., 1955, 23, 757.
45. B. Pullman. Méchanique ondulatoire et cinétique chemique, Paris, 1955, 84-108.
46. G. Wheland. J. Chem. Phys., 1949, 17, 264.
47. R. D. Brown. Trans. Far. Soc., 1948, 44, 984.
48. Anderson and Nelson. JACS, 1950, 72, 3824.
49. J. F. Tilney-Bassett and W. Waters. J. Chem. Soc., 1959, 3123.
50. M. J. Dewar. Nature, 1950, 166, 790.
51. S. H. Tucker and M. Whalley. Chem. Rev., 1952, 50, 483.
52. A. Streitwieser. Molecular Orbital Theory for Organic Chemists, N. Y., 1961.
53. M. Dewar. J. Chem. Soc., 1949, 463.
54. C. Sandorfy. Bull. Soc. Chim. France, 1949, 17, 615.
55. M. Simonetta and A. Vaciago. Nuovo cimento, 1954, 11, 596.
56. S. L. Matlow and G. W. Wheland. JACS, 1955, 77, 3653.
57. S. Basu and J. Chandhuri. Proc. Natl. Inst. Sci. India, 1958, A24, 130.
58. A. B. Almazov and D. A. Bochvar. Dokl. Akad. Nauk SSSR, 1956, 109, 121.
59. T. N. Rekasheva. Zh. Fiz. Khimii, 1955, 29, 1404.
60. P. Yvan. C. R. Acad. Sci., 1949, 229, 622.
61. C. Sandorfy and P. Yvan. Bull. Soc. Chim. Fr., 1950, 17, 131.
62. H. C. Longuet-Higgins and C. A. Coulson. Trans. Far. Soc., 1947, 43, 87.
63. P. O. Löwdin. J. Chem. Phys., 1951, 19, 1323.
64. R. Brown and Heffernan. Austral. J. Chem., 1956, 9, 83.
65. J. W. Hawarth, I. M. Heilbron and D. D. Hey. J. Chem. Soc., 1940, 349.
66. De la Mare. Chemistry and Industry, 1960, 1505.
67. R. Brown, K. Coller and K. Harcourt. Austral. J. Chem., 1961, 14, 643.
68. M. N. Adamov and I. F. Tupitsyn. Vestnik LGU, 1962, No. 22.
69. R. D. Brown and R. D. Harcourt. J. Chem. Soc., 1959, 3451.
70. R. D. Brown. Austral. J. Chem., 1955, 8, 1, 100.
71. A. Bassett. J. Chem. Soc., 1954, 2701.
72. W. Hartman and P. Sheppard. Organic Syntheses, 1942, 2, 440.
73. H. Eglish et al. JACS, 1946, 68, 453.

74. W. Davies and H. Pigott. J. Chem. Soc., 1945, 347.
75. A. Bassett and R. Brown. Chemistry and Industry, 1956, 34, 892.
76. F. Peters. J. Chem. Soc., 1958, 3763.
77. R. D. Brown. J. Chem. Soc., 1950, 690.
78. E. C. Krooman and J. Ketelaar. Rec. Trav. Chim. Pay-Bas, 1946, 65, 859.
79. R. D. Brown. Austral. J. Sci. Res., 1949, ser. A, 2, 564.
80. R. D. Brown. J. Chem. Soc., 1950, 3249.
81. Alder. Newer Methods of Preparative Organic Chemistry, 1948, 485.
82. Clar. Aromatische Kohlenwasserstoffe, 1941, 12.
83. A. Pullman and B. Pullman. Cancerisation par les substances chimiques et structure moleculaire, Paris, 1955.
84. G. Wibaut van Dijk. Rec. Trav. Chim. Pay-Bas, 1946, 65, 413.
85. G. M. Badger. J. Chem., 1949, 456.
86. F. Wallenberger. Tetrahedron Letters, 1959, 9, 5.
87. J. Meinwald. Ber., 1955, 88, 1889.
88. J. Wibaut. Chimia, 1957, 11, 228; 311.
89. F. Sixma and Wibaut. Rec. Trav. Chim. Pay-Bas, 1952, 71, 473.
90. M. J. Dewar. Records Chem. Progr., 1958, 19, 1; P. Bavin and M. Dewar. J. Chem. Soc., 1956, 164, 1441, 3570.
91. C. Hammond. JACS, 1955, 77, 334.
92. P. Norman and I. Rodda. J. Chem. Soc., 1961, 3611.
93. K. Fukui, T. Yonesava and C. Nagata. J. Chem. Phys., 1953, 21, 174.
94. K. Fukui, Tm. Yonesava. Shingu J. Chem. Phys., 1952, 20, 722.
95. H. Muller, C. Pickett and R. S. Mulliken. JACS, 1954, 76, 4770.
96. O. Chalvet. Rev. Question Scient., 1956, 17, 429.
97. R. D. Brown. J. Chem. Soc., London, 1959, 2232.
98. Nagakura. J. Chem. Phys., 1954, 22, 563.
99. Higasi and Kanite. Molecular Structure and Related Problems, Japan Sapporo, 1954.
100. K. Fukui. Bull. Chem. Soc. Jap., 1954, 27, 423.
101. M. M. Mestechkin. Vestnik LGU, 1960, No. 22.
102. M. Dewar and H. Schmeising. Tetrahedron, 1960, 11, 96.
103. H. Baba. Bull. Chem. Soc. Jap., 1947, 30, 147.
104. K. Fukui, L. Teijiro and C. Nagata. J. Chem. Phys., 1957, 26, 831.
105. T. Cottrell, The Strengths of Chemical Bonds [in Russian], Moscow, Foreign Literature Press, 1956.
106. B. Pullman and A. Pullman. Progress in Organic Chemistry, London, 1958, 4, 175.
107. A. Streitwieser. JACS, 1952, 74, 5288.

108. E. Hayashi, F. Yonezawa, G. Nagata and K. Fukul. J. Pol. Sci., 1956, 20, 537.
109. N. P. Borisova and L. P. Bokacheva. Vestnik LGU, 1963, No. 16, 138.
110. N. P. Borisova. Zh. Strukt. Khimii, 1961, 10, 113.
111. Ya. K. Syrkin. Izv. Akad. Nauk SSSR, Seriya Khimii, 1959, 238; 389; 401.
112. A. S. Onitsenko. Dienovyi sintez [Diene Synthesis], Moscow, Press of Acad. Sci. USSR, 1963.

Chapter IV

CERTAIN PROBLEMS OF INTERELECTRON AND NON-ADIABATIC INTERACTION IN LONG MOLECULES

E. E. Nikitin

The greatest difficulty encountered in modern theoretical computations of many-electron atoms and molecules lies in proper accounting for the Coulomb interaction between electrons and in the correct choice of nuclear configurations, with subsequent allowance for the non-adiabatic interaction of electronic an nuclear motions. The first problem may be solved by proper synthesis of the many-electron wave function. The methods most widely used for achieving this are based either on allowance for interactions of the nuclear configuration or on the introduction, in an explicit form, of interelectron coordinates into the test wave function. The first of these methods was largely developed by Löwdin [1], Yutsis et al. [2] and some others. These authors generalized the usual Hartree-Fock scheme and extended it to allow interaction of the configurations and utilization of open-shell functions. As for the second method, substantially new results have been obtained by Mulliken's group in Chicago [3].

The main difficulties associated with these computations are the calculation of the matrix elements between configurations and the inclusion of a large number of these elements into the computation. It would appear that new qualitative information regarding the electronic structure of molecules can only be gained by simplifying the problem. The recent successes of solid state physics in the solution of many-electron problems (e.g., [4]) are largely due to just such simplifications. The point is that in the spatially homogeneous case self-consistent MO's can be formed from mere consideration of symmetry, without solving the wave equation. Moreover, it is often possible to predict with some accuracy, even in the case of a periodic structure, the form of the self-consistent MO's, and, if we limit ourselves to the linear combination of atomic orbitals (LCAO), then the corresponding coefficients can be determined precisely. This always gives rise to the problem of how well the LCAO MO's can approximate the Hartree-Fock MO's; however, it is hoped that the correction for electron correlation and for the associated effects will prove larger than the errors introduced by the LCAO approximation, so that no new qualitative effect will be overlooked.

It is therefore of great interest to find out to what extent the methods developed in the solid state and plasma theories can be extended to the quantum mechanics of molecules. It is obvious that at the beginning we should limit ourselves to the simplest case, that is, we should investigate the effect of π-electron interactions in long conjugated molecules on the electrical and magnetic properties of these substances.

The simple one-electron theory of long conjugated molecules predicts an electron spectrum consisting of a half-filled band of width 4β, where β is the resonance integral of the C–C bond. If not all bonds are equivalent, and if their lengths alternate, this band splits into a filled and an empty one, the width of the forbidden zone being $3(\beta - \beta')$, where the quantities β and β' represent the resonance integrals of the alternating bonds. In the case of a filled band the molecule may be compared to a one-dimensional metal, and in the second case, to an insulator or a semiconductor. The excitation spectrum of such a one-dimensional metal as a function of the crystal momentum of the entire system is shown in Fig. 1 (shaded areas), while Fig. 2 represents the spectrum of a three-dimensional metal for the same one-electron excitation. The two spectra differ only in that in the three-dimensional case the spectrum is broader at small values of q, owing to the greater number of available degrees of freedom.

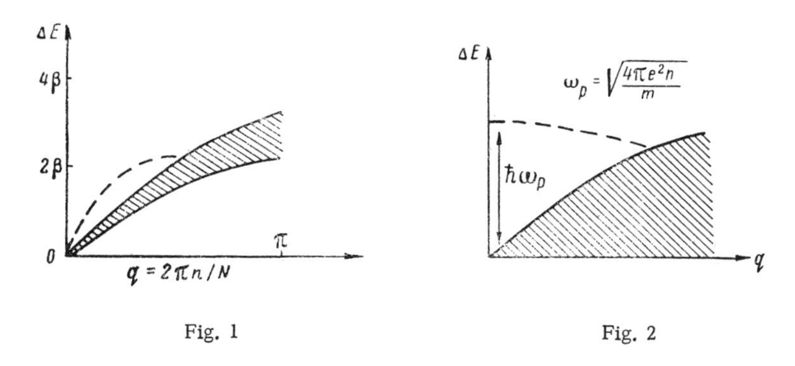

Fig. 1

Fig. 2

An important forward step in the theory of metals was the demonstration of the existence of collective vibrations of the electrons. The basic characteristic of collective vibrations is the fact that the corresponding degrees of freedom account, to a large degree, for all the electron interactions, while the remaining degrees of freedom, whose number is very much greater than the number of those corresponding to the collective vibrations, describe the motion of quasi-particles which are almost free. The frequency of these collective (so-called plasma) vibrations is equal to $\omega_r = (4\pi e^2 n/m)^{1/2}$, and the vibrations themselves may be regarded as an acoustic wave (i.e., they correspond to periodic changes in the density of the Fermi gas). As can be seen from Fig. 2, in the long-wave region ($q \ll 1$) the frequency of the vibrations is independent of the wavelength. This reflects a known property of Coulomb interactions, namely that the Coulomb attraction between planar regions of compressions and rarefactions of the Fermi gas is independent of the distance between them. This result has been obtained from methods of perturbation theory, and particularly the diagram technique. If we formally regard the interactions between electrons as a perturbation, and restrict the infinite perturbation-theory series to the sum of a definite infinite subset of terms, we can extend the resulting function of the coupling constant analytically into the region of strong interactions, and thus obtain a new perturbation approximation in terms of a new small parameter. In particular, for electrons in a metal this subset is determined by the most widely differing matrix elements of long-range Coulomb repulsion in each order, and the new small parameter is a quantity reciprocal to the electron density. Thus, a new plasma level splits off at the upper limit of the excitation energy, and, as can be seen from Fig. 2, at $q = 0$ it is the only possible one.

A problem now arises as to what extent does this picture of elementary excitations in a metal correspond to the electron excitations in conjugated molecules. On the basis of our previous discussion it is very easy to explain the qualitative dependence of the collective vibration levels on the crystal momentum. First of all, it is clear that we can neglect in the infinite series of the perturbation theory the contributions of all those excited states which are responsible for the excitation of π-electron vibrations transverse to the axis of the molecule. We can use the Coulomb potential $1/r_{12}$, averaged over the ground state of the transverse part of the one-electron MO's. For linear molecules (the only ones in which the analogy with a one-dimensional metal still holds), the mean value of $1/r_{12}$ can be represented by a function of the form

$$\left\langle \frac{1}{r_{12}} \right\rangle = \frac{1}{\sqrt{a^2 + (s_1 - s_2)^2}}, \tag{1}$$

where s_1, and s_2 are linear coordinates of electrons 1 and 2. The force of interaction between compression and rarefaction waves in the one-dimensional case will therefore decrease with increasing wavelength, so that the sepctrum of the collective vibrations should start at zero at $q = 0$ (cf. Fig. 1).

This conclusion, which can also be obtained more rigorously [5], is very important. The point is that for a number of long molecules the experimental frequency of the first absorption band does not decrease as $1/N$ (N is the number of carbons in the conjugated chain) when the length of the molecule is increased, but tends to some finite limit. It would be tempting to explain this limiting frequency by means of collective vibrations (see, for example, [6]) by analogy with the corresponding effects in metals (that is, quantized loss of energy by fast electrons passing through a foil), but the above discussion indicates that this cannot be so.

To solve one-electron problems where the one-electron spectrum is known to some degree of certainty, it is convenient to write the Hamiltonian for the system in the representation of second quantization. The usefulness of this procedure is not merely due to the automatic inclusion of Pauli's principle (even though this is very important), but also to the fact that in this new representation we can make reasonable approximations which are very difficult to formulate in the coordinate representation (especially the use of non-local interaction and of functions with undefined number of particles). Starting from given one-electron MO's we can determine the function of the ground state Φ_0 in the zero approximation, up to the Fermi limit ε_F (this function is constructed from doubly-filled

MO's). If we now define the creation operators a_k^*, b_k^* and the annihilation operators a_m, b_m for electrons and holes with respect to the level ε_F, then Φ_0 will be the vacuum function. The Hamiltonian of a system of interacting electrons is of the form

$$H = H_0 + H_C, \tag{2}$$

where H_0 is the one-electron component and H_C the Coulomb interaction of the electrons. If the problem is now solved by the method of interaction of configuration, the exact function should be sought in the form:

$$\Psi_E = \sum c_{mn\ldots;\, m'n'\ldots} a_m^* a_n^* \ldots a_{m'} a_{n'} \Phi_0. \tag{3}$$

The coefficients c must be found from a secular equation whose dimensionality is equal to the number of configurations used:

$$\frac{\partial}{\partial c} (\Psi_E^*, (H - E) \psi_E) = 0. \tag{4}$$

Different sets of the coefficients c determine the ground and excited states ψ_E. It is not difficult to see that such a solution is equivalent to the summation of certain infinite subsets of the perturbation theory series, the selection of these subsets being somewhat arbitrary. To find out which sets should be summed in the first place, it is convenient to represent operators in the Hamiltonian and terms of the perturbation theory series in the form of graphs or diagrams. The operators of creation (a_k^*) and annihilation (a_k) are represented by a vertical line beginning or ending at a point on a dotted interaction line. Thus a diagram of a form

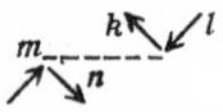

k↖↙l
m- - - - - -
↗↘n (5)

corresponds to an operator $V_{klmn} a_k^* b_l^* a_m b_n$, and a diagram of the form

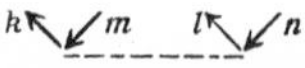

k↖↙m l↖↙n
- - - - - - - - (6)

corresponds to an operator $V_{klmn} a_k^* a_l^* b_m^* b_n^*$.

Thus, the multiplication of matrices in the energy representation is made to correspond to all possible pairings of such diagrams, with the condition that the number of free ends should be equal to the number of free indices in the product matrix. Since the magnitude of element V_{klmn} depends on the indices, and the index of the lines of electrons or holes remains constant when the diagrams are paired, it is fairly easy to sum a series of diagrams giving the maximum contribution in each order.

The use of this technique has been particularly effective in calculating the lower excitation energies in a system of interacting particles. In quantum chemistry, the excitation energy is generally calculated as a small difference between two large values (the energies of the excited and the ground state). However, the problem may be stated in another manner. We seek a function in the form of series (3), but with the difference that the known zero-approximation function Φ_0 is replaced by an as yet unknown exact function of the ground state Ψ_0. A certain equation can then be obtained for the operator $A^* = \sum c_{mn \ldots,\, m'n' \ldots} a_m^* \ldots a_{m'}$. This equation may be solved if we make certain definite assumptions regarding Ψ_0. One of these assumptions is that we can neglect the mutual interactions of electrons above the Fermi level and the interactions of holes below the Fermi level. It is clear that it would be very difficult to show such an approximation in coordinate representation, because it partially violates the Pauli principle. Assuming, in the general case, that the number of virtual pairs is small, we can write

$$\Psi_E(q) = \sum_p \left(\alpha_p(q)\, a_{p+q}^* b_p^* + \beta_p(q)\, b_{p+q} a_p\right) \Psi_0, \tag{7}$$

where α_p and β_p are parameters which will be determined. Note that if Ψ_0 is a Hartree-Fock function, then the second term in the brackets is always zero. Now consider a long chain composed of diatomic molecules. This model may be either a one-dimensional dielectric or a π-electron system in which the lengths of the C–C bonds alternate. In the zero approximation, the spectrum of such a system consists of two bands with energies $\varepsilon = \varepsilon_\nu(p)$ and $\varepsilon = \varepsilon_c(p)$. Then, the equations corresponding to the approximation (7) will be of the form

$$\begin{gathered} \left\{(E - E_0) - [\varepsilon_c(p+q) - \varepsilon_\nu(p)]\right\}\alpha_p(q) + \\ + \sum_{p'} V_1(p, p', q)\alpha_{p'}(q) - \sum_{p'} V_2(p, p', q)\beta_{p'}(q) = 0. \\ \{(E - E_0) - [\varepsilon_\nu(p+q) - \varepsilon_c(p)]\,\beta_p(q) - \\ - \sum_{p'} V_1(p, p', q)\beta_{p'}(q) + \sum_{p'} V_2(p, p', q)\,\alpha_{p'}(q) = 0. \end{gathered} \tag{8}$$

It is an important fact that only the excitation energies $\Delta E = E - E_0$ appear as eigenvalues in this system of equations. To find the significance of the matrix elements of the electron interaction, we shall limit ourselves to the Pariser-Parr-Pople approximation, neglecting all integrals in which one of the electrons occupies AO's belonging to different units of the chain. Then only integrals of the form

$$\langle cc\,|\,R\,|\,\nu'\nu'\rangle = \int \varphi_c^*(x)\,\varphi_c(x)\,\frac{1}{|x - x'|}\,\varphi_{\nu'}^*(x')\,\varphi_\nu'(x')\,dx\,dx'$$

(and consequently $\langle cc\,|\,R\,|\,c'c'\rangle$) remain, whereby c and c' refer to units separated by a distance R, and φ_ν and φ_c denote functions of the ground and excited states of the diatomic unit. For a cyclic or an infinitely long molecule we can now write

$$\begin{aligned} V_1(p, p', q) &= \sum_k \exp\,[iR(p' - p)]\,\langle cc\,|\,R\,|\,\nu'\nu'\rangle - \\ &\quad - \sum_k \exp\,[iqR]\,\langle c\nu\,|\,R\,|\,c'\nu'\rangle, \\ V_2(p, p', q) &= \sum_k \exp\,[iR(p' - p)]\,\langle c\nu\,|\,R\,|\,c'\nu'\rangle - \\ &\quad - \sum_k \exp\,[iqR]\,\langle c\nu\,|\,R\,|\,c'\nu'\rangle. \end{aligned} \tag{9}$$

Consider first the situation in which the functions of the ground and excited states of the diatomic unit, φ_v and φ_c, are not localized at various atoms (i.e., when they are not functions of polar or covalent bonds). All integrals in (9) are then of the same order of magnitude, but each sum in (9) makes a different contribution to Eqs. (8). After summation over p', the first sums in (9) can, under certain conditions, make a small contribution, owing to oscillation of the factor $\exp[iR(p'-p)]$. We can then write $V_1(p, p', q) \sim V_2(p, p', q) \approx J(q)$, and Eqs. (8) will have the form

$$\begin{aligned}
&\left\{\Delta E - \left[\varepsilon_c(p+q) - \varepsilon_v(p)\right]\right\} \alpha_p(q) - \\
&\quad - J(q) \sum_{p'} \left[\alpha_{p'}(q) - \beta_{p'}(q)\right] = 0, \\
&\left\{\Delta E - \left[\varepsilon_v(p+q) - \varepsilon_c(p)\right]\right\} \beta_p(q) + \\
&\quad + J(q) \sum_{p'} \left[\beta_{p'}(q) - \alpha_{p'}(q)\right] = 0.
\end{aligned} \tag{10}$$

Solutions of these equations for the one- and three-dimensional cases are given in Figs. 1 and 2. If we had taken Ψ_0 to be a Hartree-Fock function from the beginning, then we would not obtain a new excitation branch. The above approximations imply that in the perturbation-theory graphs we must sum up matrix elements which correspond to the same transmitted momentum q, considering only the electron-hole interactions, i.e., diagrams of the type

$$\Lambda\text{---}0\text{---}0\text{---}V.$$

The significance of the collective level is easily explained using an H_2 molecule as an example. We shall consider as our basis functions two molecular orbitals σ_g and σ_u, which are self-consistent in this limited set of functions. The Hartree-Fock function of the ground state Φ_0 corresponds to the configuration σ_g^2. At large internuclear distances R, this function is known to represent the actual facts very poorly. If we take into account the interaction between configurations σ_g^2 and σ_u^2 we can find a more exact function, which can be represented as a linear combination of Heitler-London function Ψ and ionic-structure function $\Psi(I)$, i.e., $\Psi_0 = \Psi(HL) + \lambda'\Psi(I) = \sigma_g^2 + \lambda\sigma_u^2$. As $R \to \infty$, $\lambda \to 1$, and $\lambda' \to 0$.

It is not difficult to see that the operator for the transition from the ground state function to the excited state function Ψ_E is easier to find than the functions proper. In this case, the function of the first excited state Ψ_E can be obtained from $\Psi_0 = \sigma_g^2 + \lambda\sigma_u^2$ by the action of operator P_{gu} which permutes indices g and u. Operator P_{gu} which substitutes u for g corresponds in secondary quantization representation to the formation of a hole and an electron, while substitution of g by u denotes pair annihilation. During this the parameter λ drops out completely. If we now substitute the function $\Psi_E = P_{gu}\Psi_0$ into the Schrödinger equation, we would find the correct excitation energy, which otherwise could only be obtained by calculating the interaction of configurations σ_g^2 and σ_u^2. Note that we could not improve the situation by applying the operator P_{gu} to Φ_0. Comparing this conclusion with previous considerations, it is fairly easy to see that such a determination of Ψ_0 and Ψ_E should take electron correlation into account. Considering the case which is most unfavorable for the application of the Hartree-Fock function, that of large R, we find that three singlet states can be represented approximately by the Heitler-London functions $\Psi(HL) = \sigma_g^2 - \sigma_u^2$ and by two ionic structures, $\sigma_g\sigma_u\left({}^1\Sigma_u^+\right)$ and $\sigma_g^2 + \sigma_u^2\left({}^1\Sigma_g^+\right)$. From this we see that excited states do, in fact, correspond to vibrations of the electron plane. Conversely, as a result of electron correlation, energetically unfavorable ionic structures are separated out from the ground state.

Let us now consider a second case, in which the functions φ_v and φ_c overlap to a small extent. In the one-electron approximation, this leads to the appearance of a forbidden zone between the valence and the conduction bands. Under these conditions the integrals $\langle c\nu | R | c'\nu' \rangle$ may prove to be small in comparison with $\langle cc | R | \nu\nu' \rangle$, and we can assume that approximately

$$V_1(p, p', q) = \sum \exp[iR(p'-p)] \langle cc | R | \nu'\nu' \rangle,$$
$$V_2 \approx 0,$$

so that Eqs. (8) assume the form

$$[\Delta E - \varepsilon_c(p+q) + \varepsilon_\nu(p)]\, \alpha_p(q) + \\ + \sum_{k,\, p'} \exp[iR(p'-p)] \langle cc | R | \nu'\nu' \rangle\, \alpha_{p'}(q) = 0, \quad \beta_p(q) = 0. \tag{11}$$

The corresponding excitation spectrum at small q (Fig. 3) shows that the collective level lies below the spectrum of one-particle excitations. Transforming (11) into coordinate representation we can readily see that it corresponds to a coupled electron-hole state. Such an exciton excitation could also have been obtained by making $\Psi_0 = \Phi_0$ at the very beginning. Linear combination of singly excited Hartree-Fock configurations therefore provides a satisfactory description of the excited states. The solution of Eqs. (11) is therefore equivalent to the summation of diagrams which contain the maximum number of non-annihilated electrons and holes at each order of perturbation theory, i.e., diagrams of the form

In actual practice, when hardly any of these approximations are completely valid, we cannot limit ourselves exclusively to plasmon or exciton diagrams, but must also consider their interaction. We could obtain satisfactory separation of the two types of excitation had the exchange and Coulomb integrals between the Hartree-Fock MO's been very different. Unfortunately, the situation is worse in the one-dimensional case than in the three-dimensional. The reason is that the Coulomb interaction of electrons in long

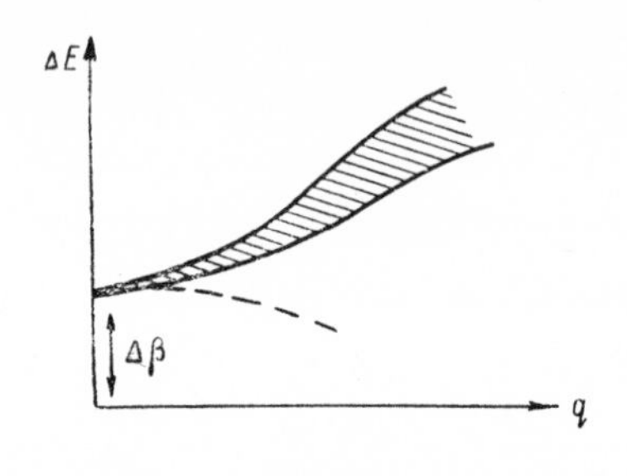

Fig. 3

molecules is neither long-range nor short-range, so that it is difficult to distinguish those terms of the entire perturbation theory series which make the principal contributions. This is also reflected by the fact that, on transition from the three-dimensional to the one-dimensional case, the usual $1/q^2$ relation of Coulomb interaction matrix elements is replaced by $\ln q$.

We could hardly expect the summation of terms of the perturbation theory series to account in a reasonably satisfactory manner for electron interaction under these conditions. The problem might perhaps be phrased somewhat differently; thus we can introduce some model for interelectron interaction. This would permit us a solution of the many-particle problem to some good approximation. We could then find out whether such a treatment of the electron interaction does indeed lead to new qualitative conclusions.

In particular, let us consider the case of large cyclic molecules. In these, electron interaction is a periodic function of the angle describing the distance between electrons. We could retain only the first two terms out of the overall expansion

$$\left\langle \frac{1}{r_{12}} \right\rangle = 2A_0 + 2A_1 \cos \varphi_{12} + 2A_2 \cos 2\varphi_{12} + \dots \tag{12}$$

In the approximation (7) considered above, the Hamiltonian of such a system coincides (with an accuracy to terms of the order of $1/N$, where N is the number of π-electrons) with the Hamiltonian of N interacting oscillators. Since the latter Hamiltonian can be solved exactly, it is not difficult to determine the coefficients c_k of the expansion of a function of interacting oscillators in terms of functions of a system without interaction. But this also means that, to an accuracy of $1/N$, the c_k can be regarded as coefficients of the expansion of the functions of interacting π-electrons in functions of noninteracting electrons. Thus, an estimate of the contribution of the excited configurations $\Phi_0^{(k)}$ to the function of the ground state Ψ_0 gives

$$|c_k|^2 = \left| \int \Psi_0^* \Phi_0^{(k)} d\tau \right|^2 = \frac{2\beta}{1+\beta} \frac{\Gamma\left(m+\frac{1}{2}\right)}{\Gamma(m+1)} \left(\frac{1-\beta}{1+\beta}\right)^{2m} \text{ for } k=2m; \quad (13)$$
$$|c_k|^2 \text{ for } k=2m+1,$$

where $\beta = (1+4A_1/\Delta\varepsilon)^{1/2}$ and $\Delta\varepsilon$ is the energy of the first excitation, calculated by the simple LCAO method or the metallic model. The energies of excitation of the four lowest states $M = 1, N/2$, $S = 0.1$ (which are degenerate in the absence of interaction) are equal to

$$\Delta\varepsilon \cdot \beta \left(\Psi_0 \rightarrow \Psi_{M=1}^{S=0}\right) \text{ and } \Delta\varepsilon\left(\Psi_0 \rightarrow \Psi_{M=1}^{S=1},\ \Psi_{M=\frac{N}{2}}^{S=0,1}\right).$$

The usual condition of perturbation theory that the matrix element of the interaction be small in comparison with the excitation energy is replaced here by the condition that it be small in comparison with the Fermi energy. On the basis of this model, it is easy to see the meaning of the high-density approximation in the calculation of collective effect in a one-dimensional Fermi gas. The accuracy of such calculations increases with the total number of electrons. In real molecules, however, N increases in proportion to the radius of the ring of the molecule and the higher terms in expansion (12) which were neglected begin to play a more and more important part.

Our discussion of the above problems was made within the limits of the adiabatic approximation. However, even in this case it is very difficult to determine the equilibrium configuration of the nuclei. The difficulty is connected with the fact that, in the approximate variational selection of the electron functions, restrictions are very often placed on the movements of nuclei, which make it impossible to find the true equilibrium configuration by variation of the internuclear distances. Moreover, after we find a solution, the adiabatic approximation we used may turn out to be inapplicable. A typical example of this is the dynamical effect of Jahn-Teller in electronically excited cyclic molecules and ion radicals. The splitting of a degenerate electronic state as a result of a distortion in the symmetrical configuration of nuclei is of the same order of magnitude as vibrational quanta. It is clear that we cannot regard the nuclei as localized in the new equilibrium positions, and the wave function must therefore be sought in the form of a general expansion

$$\Psi = \sum_{n, m} \Psi^{n}_{\text{el.}}(R, r_{\text{el.}}) \chi^{nm}_{\text{vib.}}(R), \tag{14}$$

in which each term describes a certain electronic vibrational state in the adiabatic approximation. When the non-adiabatic interaction is considerable, it is not, of course, desirable to use the functions $\Psi^{n}_{\text{el.}}(R, r_{\text{el.}})$ as the zeroth approximation, but we can limit ourselves to the electronic functions for a fixed configuration $R = R_0$. A certain analogy with the Hartree-Fock and Hartree-Fock CI approximations (where CI indicates configuration interaction) can now be drawn. If we consider only one configuration, we must use the one-electron Hartree-Fock MO's as the basis. The more configurations are considered, the less decisive becomes the selection of the basis.

At the present time, systems of equations determining the vibrational functions in (14) have been set up for the most important examples of the Jahn-Teller dynamic effect. The eigenvalues have also been determined approximately.

The appearance of a forbidden zone in the spectra of long molecules containing conjugated bonds is very closely connected with the Jahn-Teller effect. Any distortion of the structure of the symmetric chain leads to the splitting of a half-filled band into a series of sub-bands, i.e., to the appearance of forbidden zones. Only one of all possible modes of deformation leads to a reduction of the energy of the system of x-electrons. This deformation corresponds to alternate compression and extension of C-C bonds, i.e., alternations of their lengths. In the adiabatic one-electron approximation, the total energy of the molecule is composed of the energy of the π-electrons, E_π, and the energy of a deformed σ-framework, E_σ. Considering only the single type of deformation Q, shown above we can determine the minimum of $E = E_\pi + E_\sigma$. Clearly, the energy deformation of the σ-framework for a given difference in the lengths of the alternating bonds is proportional to N. On the other hand, it can be shown that $E_\pi = \varepsilon_\pi N \ln N$ [7].

It is thus seen that as the length of a molecule is increased, a point is always reached at which the symmetry of the configuration becomes useful. The corresponding critical magnitudes of N have been determined by Longuet-Higgins et al. [7] and found to be of the order of 20-30.

An important qualitative conclusion, which follows from the above and which is illustrated in Fig. 4, is that the total gain of energy

ΔE originating during deformation may be considerably lower than the energy ΔE_{π} characterizing the increase of frequency in the π-electron spectrum. For example, even in the relatively small molecule of pentalene the total reduction of the energy is 0.14 β, while the energy of the first excitation varies between 0.11β and 0.52β. Thus, we cannot even discuss equilibrium configuration unless ΔE is considerably greater than the frequency of the vibrational quantum corresponding to the normal vibration Q. However, for the general case of known long molecules, we cannot assert that this happens. In other words, to describe the spectrum of such molecules we must start from the full electronic and vibrational Hamiltonian.

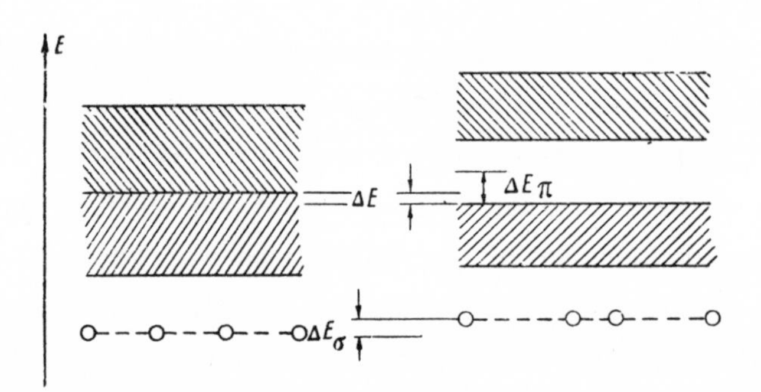

Fig. 4

In the simplest case where our molecules are cyclic and where the resonance integrals are calculated only between neighboring atoms, the Hamiltonian for one-electron MO's is of the following form in the adiabatic approximation

$$H = \sum \alpha A_r^* A_r + \sum \beta A_r^* A_{r\pm 1} + \\ + \Delta\beta \sum (-1)^n \left(A_r^* A_{r+1} - A_r^* A_{r-1} \right), \tag{15}$$

where A_r are the LCAO coefficients. The last term takes into account the alternation of the bonds. The transition to the many-electron case is achieved by replacing the coefficients A_r by the corresponding operators of single electron annihilation in the p_rth AO of the rth carbon atom. In order to take non-adiabatic conditions into account, one should take into account all the normal vibrations and the dynamic character of the interaction of electrons with lattice vibrations. Thus we go from one representation to another with the aid of transformation from the AO's to the MO's, and we obtain (in units of $2\beta_0$)

$$H = \sum (1 - \cos k)\, a_k^* a_k + \sum \Omega(p) \left[b_p^* b_p + \frac{1}{2} \right] + H',$$
$$H' = i g_1 \sum \operatorname{sign} p \exp\left(is - \frac{ip}{2} \right) \times \quad (16)$$
$$\times \sqrt{\frac{\Omega(p)}{N}}\, a_{s+p}^* a_s \left[b_p + b_{-p}^* \right] + \text{к. с.},$$

where a_k^* is the creation operator of an electron in the kth MO, b_p^* is the creation operator of a phonon

$$g_1 = \sqrt{\beta_0 \frac{\alpha^2}{2m\omega_0^2}}, \quad \alpha = \frac{1}{\beta} \frac{\partial \beta}{\partial R} \bigg|_{C-C},$$

and m is the mass of the nucleus. The advantage of writing this in form (16) is that from the very beginning the problem is one of determination of electron-phonon ground and excited states and not of an electronic state in a deformed lattice. Moreover, there exist well-developed methods for the approximate calculation of the ground state and elementary excitations of similar and even more general Hamiltonians (including electron interaction). These involve the application of methods of the theory of superconductivity to the calculation of non-adiabatic interactions in long molecules. The point is that the formation of an energy gap in the spectrum of a metal as a result of electron-phonon interaction (a gap which is responsible for the superconductivity) is to some extent analogous to the formation of a forbidden zone in a π-electron spectrum of a long molecule. Bogolyubov's method [8], based on the introduction of new operators for the creation and annihilation of quasi-particles, which interchanges electrons and holes, allows approximately for the interaction with any phonon, whereas the LCAO treatment using the adiabatic Hamiltonian (15) is restricted to the phonon of the shortest wavelength. Furthermore, Bogolyubov's method also allows us to take into account the interaction of the electrons. For this, of course, it is necessary to take into account the specific nature of the one-dimensional Coulomb interaction.

In large systems, it is much more essential to allow simultaneously for non-adiabatic and electron interactions than it would seem at first sight. If the terms for the Coulomb interaction of electrons are rejected, even in the starting Hamiltonian, the non-adiabatic wave function must necessarily be a many-electron one. This is easily seen from the fact that in higher orders of perturbation theory (beginning with the second) the non-adiabaticity operators will couple not only the functions of one and the same electron, but also functions of different electrons. In the formalism of secondary quantization, such a situation can be reduced to an exchange of electrons with virtual phonons. It is particularly important that the probability of the absorption of a phonon by other electrons is approximately N times greater than the probability of its absorption by the electron which emitted it. It is, indeed, the allowance for this phenomenon which permitted the energy gap in superconductors to be calculated to the correct order of magnitude. It is interesting to note that, even if we neglect the Coulomb interaction of electrons, the approximate variational wave function of a superconductor has the form [9]

$$\Psi_0 = \prod_k \left(u_k + v_k a^*_{k\sigma} a^*_{-k,\,-\sigma} \right) \Phi_0. \tag{17}$$

This many-electron function does not correspond to any definite number of particles, and cannot therefore be written in the coordinate representation.

The problem of the interaction of electrons via virtual phonons in molecules has not, as yet, been studied, even though this interaction is an effect of second (and higher) order, and a number of such higher order effects have been thoroughly investigated in quantum chemistry (e.g., the indirect interaction of nuclear spins via the spin of an electron). In connection with this, it would be interesting to discuss the influence of this effect on spin-lattice relaxation in symmetrical ion radicals. So far, the relationship between electron spin and the motion of nuclei, a relationship resulting from the dynamic Jahn-Teller effect, has been considered in the one-electron approximation. It is not clear whether one really can neglect other equivalent electrons with pairwise-compensated spins.

It is thus seen that, for long molecules and molecules with orbital degeneracies, the problem of electron interaction cannot be solved without accounting for the non-adiabaticity. In these cases it may be found that the one-electron approximation is not even a zero-approximation, and that the problem should be solved from the very outset by allowing or the static and dynamic interactions of electrons and nuclei. Of course, in the case of molecules, the electron interactions can and should be calculated in the adiabatic approximation. However, qualitatively new results can evidently be obtained only for large molecules, where it is perfectly natural to apply methods developed in modern many-particle theory.

REFERENCES

1. P. O. Löwdin. Adv. in Chem. Phys., 1959, 2, 207.
2. Ya. I. Vizbaraite, T. D. Strotskite and A. P. Yutsis. Dokl. Akad. Nauk SSSR, 1961, 61.

3. Technical Report, University of Chicago, Group of Molecular Physics, 1960.
4. Voprosy kvantovoi teorii mnogikh tel [The Quantum Theory of Many-body Interactions], Collection of papers, State Press for Foreign Literature, Moscow, 1959.
5. Y. Miruno and T. Izuyama. Progr. Theoret. Phys., 1959, 21, 593.
6. G. Araki and T. Murai. Progr. Theoret. Phys., 1957, 8, 649.
7. L. Salem and H. C. Longuet-Higgins. Proc. Roy. Soc., 1959, A251, 172.
8. N. N. Bogolyubov, V. V. Tolmachev and D. V. Shirkov. Novyi metod v teorii sverkhprovodimosti [A New Method in the Theory of Superconductivity], Press of Acad. Sci. USSR, 1958.
9. Teoriya sverkhprovodimosti [Theory of Superconductivity], Collection of Problems, State Press for Foreign Literature, Moscow, 1960.

Chapter V

APPLICATION OF THE METHOD OF FINITE DIFFERENCES TO PROBLEMS OF MOLECULAR ORBITAL THEORY

T. K. Rebane

INTRODUCTION

The simple molecular orbital (MO) theory [1-3], which does not explicitly take into account the electrostatic interaction between electrons, has been widely used in the quantum-chemical studies on the qualitative and semi-quantitative characteristics of the electronic structure of molecules. The mathematical machinery of the MO method is based on matrix algebra. When the molecule contains a large number of atoms, then the use of these mathematics requires the calculation of determinants of high order, as well as the solution of algebraic equations of high order. However, in the case of a molecule possessing a regular structure, there exists a more efficient method for calculating the molecular orbitals and energy levels: it is based on the use of the simplest theory of finite difference equations with constant coefficients. Since no systematic exposition of this method has as yet appeared in the literature on quantum chemistry, it was thought desirable to outline the method in brief and to illustrate its possibilities with typical examples.

The MO method is summarized in Section 1, and the fundamentals of finite difference equations with constant coefficients are set out in Section 2. Sections 3-8 illustrate the applications of the method of finite differences to various sroblems of the MO theory.

1. A BRIEF OUTLINE OF THE FUNDAMENTALS OF THE MO METHOD

In the MO method, approximate one-electron wave functions describing the electrons in a molecule are constructed in the form of linear combinations of the wave functions of individual atoms

$$\psi_p = \sum_{k=1}^{N} c_{kp}\xi_k, \tag{1.1}$$

where ξ_k is the wave function of an electron in the kth atom and the c_{kp} are the unknown coefficients. The first index ($k = 1,2,\ldots, N$) in c_{kp} indicates the number of the atom, and the second ($p = 1, 2, \ldots, N$) the number of the MO. The quantities c_{kp} are called orbital coefficients. Since the wave functions of the various atoms are linearly independent, the number of linearly independent MO's is equal to the number of atoms, i.e., to N.

We shall limit ourselves to the simple variant of the MO theory, in which the electrostatic interaction between electrons is not explicitly taken into account. In this case the operator of the total energy of electrons in the molecule is the sum of operators $\hat{H}_i$ for individual electrons:

$$\hat{H} = \sum_{i=1}^{M} \hat{H}_i = \sum_{i=1}^{M} \left[-\frac{h^2}{2m} \Delta_i + V\left(\vec{r}_i\right) \right]. \tag{1.2}$$

Here, $h^2\Delta_i/2m$ is the kinetic energy operator for the ith electron and

$V(\vec{r}_i)$ is the potential energy of that electron which in this approximation depends only on the set of coordinates of the ith electron. The mathematics of the MO method includes two matrices of primary importance: the matrix of the integrals of non-orthogonality of functions ξ_k:

$$S_{kl} = \int \xi_k^*(\vec{r})\, \xi_l(\vec{r})\, d\tau = S_{lk}^* \tag{1.3}$$

and the matrix of the one-electron energy operator

$$H_{kl} = \int \xi_k^*(\vec{r}_i)\, \hat{H}_i \xi_l(\vec{r}_i)\, d\tau_i = H_{lk}^*. \tag{1.4}$$

where an asterisk denotes complex conjugation.

The MO's of the electrons are subject to the conditions of normalization and orthogonality

$$\int \psi_p^* \psi_q\, d\tau = \sum_{k=1}^{N} c_{kp}^* c_{kq} + \sum_{l \neq k}^{N} c_{kp}^* S_{kl} c_{lq} = \delta_{pq}. \tag{1.5}$$

Equation (1.5) takes into account the condition of normalization of the atomic wave functions

$$S_{kk} = \int \xi_k^*(\vec{r})\, \xi_k(\vec{r})\, d\tau = 1. \tag{1.6}$$

The orbital coefficients and energies of the MO's are determined from the requirement that the mean energy of an electron in a molecule, calculated with a wave function of the form (1.1) under additional conditions (1.5), be an extremum. This requirement leads to a system of N linear homogeneous algebraic equations for the c_{kp}

$$\sum_{l=1}^{N} (H_{kl} - \varepsilon_p S_{kl})\, c_{lp} = 0, \quad k,\ p = 1,\ 2,\ \ldots,\ N. \tag{1.7}$$

The ε_p denote possible values of the energy, determined from the condition that system (1.7) have a non-trivial solution. The standard method of finding ε_p consists in equating the determinant of (1.7) to zero:

$$\det(H - \varepsilon S) = 0. \tag{1.8}$$

After the ε_p are found by solving (1.8), the corresponding orbital coefficients are constructed by solving (1.7). These equations, and the condition of normalization of the MO's [obtained from (1.5) at $p=q$] uniquely determine the MO's (more precisely, they determine it to within a physically unimportant constant coefficient whose modulus is equal to unity). An exception is the case of degeneracy when two or more MO's correspond to one value of the energy: in that case the MO's are then determined to within linear orthogonal transformations which do not affect (1.5).

It will be shown below that, for molecules having a regular structure, it is preferable to use the method of finite differences to determine the energy values and to construct the MO's (see. Sections 3-8).

Different assumptions regarding the form of matrices S (1.3) and H (1.4) correspond to different modifications of the MO method. Most frequently, we consider only the interaction of atoms of the molecule which are connected by direct chemical bonds; the non-diagonal elements of S and H are assumed to be non-zero only for connected atoms k and l:

$$\left.\begin{aligned} \text{at } k \neq l \; S_{kl} &\neq 0, \\ H_{kl} &\neq 0, \end{aligned}\right\} \text{only for mutally connected atoms } k \text{ and } l \qquad \begin{aligned}(1.9)\\(1.10)\end{aligned}$$

The non-diagonal elements of H in Eq. (1.10) are usually called "resonance integrals." They are denoted by

$$H_{kl} \equiv \beta_{kl}. \tag{1.11}$$

Since the non-diagonal elements of S are, as a rule, considerably smaller than one, an even simpler form of the matrix S is often used

$$S_{kl}=\delta_{kl}. \tag{1.12}$$

The diagonal elements of matrix H are called Coulomb integrals and are denoted by α:

$$H_{kk}\equiv\alpha_k. \tag{1.13}$$

Unless otherwise indicated, we shall be using the simplest variant of the MO method, which involves the following assumptions:

1. The wave functions of various atoms are regarded as orthogonal, i.e., Eq. (1.12) is used.

2. The integrals β for all bonds in the molecule are assumed to be equal. The bonds are sometimes divided into several groups, and to each group is assigned a value of β which is the same for all bonds in the group.

3. In a given molecule, the Coulomb integrals α of all atoms of the same type are assumed to be equal.

The system (1.7) then becomes

$$(\alpha_k-\varepsilon_p)\,c_{kp}+\sum_l\beta_{kl}c_{lp}=0,\quad k,\ p=1,\ 2,\ \ldots,\ N. \tag{1.14}$$

Summation over l is then carried out only over atoms coupled to atom k. (Further details of the MO method are given in [4].)

2. SOLUTION OF A LINEAR HOMOGENOUS EQUATION IN FINITE DIFFERENCES WITH CONSTANT COEFFICIENTS

An equation of the form

$$a_nf(k+n)+a_{n-1}f(k+n-1)+\ldots+a_1f(k+1)+ \\ +a_0f(k)=0, \tag{2.1}$$

where $f(k)$ is an unknown function of integral argument is called a linear homeogeneous finite difference equation of order k. The general solution of Eq. (2.1) contains n arbitrary constants and is a linear combination of its n linearly independent particular solutions $f_l(k)$:

$$f(k)=\sum_{l=1}^{n} C_l f_l(k), \tag{2.2}$$

where

$$f_l(k)=(\lambda_l)^k. \tag{2.3}$$

The numbers λ_l ($l=1, 2, \ldots, n$) are the roots of the so-called characteristic equation, an nth order algebraic equation:

$$a_n(\lambda_l)^n+a_{n-1}(\lambda_l)^{n-1}+\ldots+a_1\lambda_l+a_0=0. \tag{2.4}$$

Note that the trivial solution of (2.1), $f(k)=0$, is not one of the linearly independent solutions.

In problems involving the MO method, it is often desirable to write λ_l in the form

$$\lambda_l=e^{i\mu_l}. \tag{2.5}$$

(Note that λ_l and μ_l can be complex.) In this notation, the lth particular solution of (2.3) is of the form

$$f_l(k)=e^{ik\mu_l}, \tag{2.6}$$

and the general solution of (2.1) becomes

$$f(k)=\sum_{l=1}^{n} C_l e^{ik\mu_l}. \tag{2.7}$$

Let us now consider the case in which the order n of Eq. (2.1) is even and the coefficients of this equation are symmetrical:

$$a_{n-l}=a_l, \quad l=0, 1, \ldots, n/2. \tag{2.8}$$

For each root, λ_l, of the characteristic Eq. (2.3) we now have another root, equal to $1/\lambda_l$. In the notation of (2.5), this pair of roots is represented by

$$\lambda_l = e^{i\mu_l}; \quad 1/\lambda_l = e^{-i\mu_l}; \tag{2.9}$$

and to this pair correspond the particular solutions

$$f_{\mu_l}(k) = e^{ik\mu_l} \text{ and } f_{-\mu_l}(k) = e^{-ik\mu_l}. \tag{2.10}$$

Since linear combinations of particular solutions of (2.1) are also solutions, solutions (2.10) can be replaced by

$$\begin{aligned} \varphi_l(k) &= \frac{1}{2}\left[f_{\mu_l}(k) + f_{-\mu_l}(k)\right] = \cos k\mu_l; \\ \chi_l(k) &= \frac{1}{2i}\left[f_{\mu_l}(k) - f_{-\mu_l}(k)\right] = \sin k\mu_l. \end{aligned} \tag{2.11}$$

It follows that the general solution of a linear homogenous equation in finite differences with constant coefficients, with n even, can be written in the form

$$f(k) = \sum_{l=1}^{n/2} (A_l \cos k\mu_l + B_l \sin k\mu_l). \tag{2.12}$$

The above outline is sufficient for the solution of a number of problems in MO theory. A more rigorous and detailed exposition of this theory can be found in specialized literature, for example, in the book [5].

3. MOLECULAR ORBITALS AND ENERGY LEVELS FOR CHAINS AND RINGS COMPOSED OF IDENTICAL ATOMS WHEREBY ALL THE BONDS ARE FULLY EQUALIZED

As a first example of the application of the theory of finite difference equations, we shall construct the MO for a chain of N identical atoms having equal Coulomb integrals α. The bonds in this chain are regarded as fully equalized, so that all resonance integrals are equal to β.

In this case, Eq. (1.14) for the orbital coefficients has the form

$$\beta c_{k-1} + (\alpha - \varepsilon) c_k + \beta c_{k+1} = 0, \quad k = 1, 2, 3, \ldots, N. \tag{3.1}$$

The boundary conditions

$$c_0 = 0; \quad c_{N+1} = 0. \tag{3.2}$$

should take into account the absence of atoms at the end of the chain. These conditions ensure that (3.1) is correct at $k = 1$ and $k = N$.

The set of equations given by (3.1) is a linear second order homogeneous equation in finite differences with constant, symmetrical coefficients for the function c_k of integral argument k. According to (2.12), the general solution of (3.1) is:

$$c_k = A_\mu \cos k\mu + B_\mu \sin k\mu. \tag{3.3}$$

From our first boundary condition, we have $A_\mu = 0$. Since A_μ and B_μ cannot both be zero at the same time, the second boundary condition of (3.2) gives

$$\sin (N+1)\mu = 0. \tag{3.4}$$

Therefore, the possible values of the parameter μ are

$$\mu = p\pi / (N+1). \tag{3.5}$$

Different values of the integer p correspond to different MO's. The p-values, $p = 1, 2, \ldots, N$ correspond to N linearly independent MO's.

According to (3.3), and the fact that $A_\mu = 0$, the orbital coefficients of the pth MO are

$$c_{kp} = B_p \sin \frac{kp\pi}{N+1}. \tag{3.6}$$

To determine the energy of the MO, we substitute (3.6) into (3.1), and obtain a condition which should be fulfilled at all values of

$k\,(k=1,\,2,\,\ldots,\,N)$:

$$B_p \sin\frac{kp\pi}{N+1}\left[\alpha-\varepsilon+2\beta\cos\frac{p\pi}{N+1}\right]=0. \tag{3.7}$$

Hence the expression for the energy ε_p of the pth MO is

$$\varepsilon_p=\alpha+2\beta\cos\frac{p\pi}{N+1}. \tag{3.8}$$

The factor B_p is determined from the normalization condition [see (1.5)] by assuming $S_{kl}=\delta_{kl}$ [see (1.12)]

$$1=\sum_{k=1}^{N} c_{kp}^2=B_p^2\sum_{k=1}^{N}\sin^2\frac{kp\pi}{N+1}=\frac{N+1}{2}B_p^2. \tag{3.9}$$

The normalized orbital coefficients are then obtained in the form

$$c_{kp}=\sqrt{\frac{2}{N+1}}\sin\frac{kp\pi}{N+1}. \tag{3.10}$$

The problem is thus completely solved for a chainlike molecule.

We shall now consider an analogous problem for a cyclic molecule consisting of N atoms. As before, the orbital coefficients satisfy Eqs. (3.1), but the boundary conditions (3.2) must now be replaced by periodicity conditions which ensure that the c_k are unique.

$$c_{k+N}=c_k. \tag{3.11}$$

It is convenient to seek the solution of Eq. (3.1) with conditions (3.11) in the form of an exponential function

$$c_k=C_\mu e^{ik\mu}. \tag{3.12}$$

The possible values of the parameter μ follow from (3.11):

$$\mu=\frac{2p\pi}{N}. \tag{3.13}$$

Here p is an integer. When N is even, the linearly independent orbital coefficients correspond to the following values of p:

$$p=0, \pm 1, \pm 2, \ldots, \pm\left(\frac{N}{2}-1\right), \frac{N}{2}; \tag{3.14}$$

and, when N is odd, to

$$p=0, \pm 1, \pm 2, \ldots, \pm \frac{N-1}{2}. \tag{3.15}$$

Substituting (3.12) into (3.1), we obtain for the energy [see the derivation of Eqs. (3.7) and (3.8)]:

$$\varepsilon_p=\alpha+2\beta \cos \frac{2p\pi}{N}. \tag{3.16}$$

This equation shows that the energy depends only on the absolute value of p. Energy levels corresponding to $p=\pm 1, \pm 2, \ldots, \pm (N/2-1)$ (with N even) and to $p=\pm 1, \pm 2, \ldots, \pm (N-1)/2$ (with N odd) are therefore doubly degenerate. The level $p=0$ is not degenerate either at even or odd N, and the level $p=N/2$ is not degenerate at even N.

The orbital coefficients given by (3.12) are readily normalized. We have

$$c_{kp}=\frac{1}{\sqrt{N}} e^{\frac{2ikp\pi}{N}}. \tag{3.17}$$

Consider N to be even and the number of electrons M to be equal to the number of atoms ($M=N$). The lowest $N/2$ levels will then be occupied in the ground electronic state of the molecule. The first electronic transition (long-wave absorption band) corresponds to the transition of an electron from the last occupied level to the first free level. From (3.8) and (3.16), it follows that the energy of this transition for a chain is

$$\Delta\varepsilon(N)=-4\beta \sin \frac{\pi}{2(N+1)}, \quad \beta<0, \tag{3.18}$$

while for a ring with odd $N/2$ it is

$$\Delta\varepsilon(N) = -4\beta \sin\frac{\pi}{N}, \quad \beta < 0. \tag{3.19}$$

These equations show that, in chain and ring molecules consisting of identical atoms and having fully equalized bonds, the electronic excitation energy tends to zero as the number of atoms $N \to \infty$.

4. ENERGY LEVELS FOR A RING AND A CHAIN WHEN THE VALUES OF THE COULOMB AND RESONANCE INTEGRALS ALTERNATE

Consider now a molecule in which two different types of atoms (characterized by Coulomb integrals α_1 and α_2) alternate. We shall assume that the bonds between these atoms are not quite equalized, so that the resonance integrals for two adjacent bonds are alternately β_1 and β_2. A fragment of such a molecule is shown in Fig. 1.

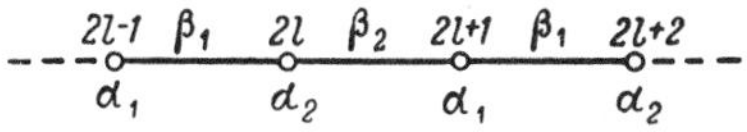

Fig. 1

The Coulomb integrals of atoms with odd indices shall be denoted by α_1 and those of atoms with even indices by α_2.

In this case, Eqs. (1.14) are of the form

$$\begin{aligned} \beta_1 c_{2l-1} + (\alpha_2 - \varepsilon) c_{2l} + \beta_2 c_{2l+1} &= 0, \\ \beta_2 c_{2l} + (\alpha_1 - \varepsilon) c_{2l+1} + \beta_1 c_{2l+2} &= 0. \end{aligned} \tag{4.1}$$

We exclude from this equation, coefficients c_{2l} (where l is an integer) and obtain

$$\beta_1\beta_2 d_{l-1} + [\beta_1^2 + \beta_2^2 - (\varepsilon - \alpha_1)(\varepsilon - \alpha_2)]\, d_l + \beta_1\beta_2 d_{l+1} = 0, \tag{4.2}$$

where

$$d_l \equiv c_{2l+1}. \tag{4.3}$$

Equations (4.2) are equivalent to a second-order linear homogeneous equation in finite differences with constant, symmetrical coefficients.

For the case of a cyclic molecule containing N atoms (N being even) with indices $k=1, 2, \ldots, N$, the periodicity condition

$$c_{k+N}=c_k, \tag{4.4}$$

must be fulfilled. On changing to coefficients d_l, this is written as

$$d_{l+N/2}=d_l. \tag{4.5}$$

As in Section 3, we seek the solution of Eq. (4.2) in the form

$$d_l=C_\mu e^{il\mu}. \tag{4.6}$$

Repeating the operations carried out in Section 3 (with the sole difference that we replace N by $N/2$), we obtain $N/2$ linearly independent solutions of (4.2)

$$d_{lp}=C_p e^{\frac{4lp\pi i}{N}}, \tag{4.7}$$

where for $N/2$ even, $p = 0, \pm 1, \pm 2, \ldots, \pm$

$$\pm\left(\frac{N}{4}-1\right), \frac{N}{4}; \tag{4.8}$$

and for $N/2$ odd,

$$p=0, \pm 1, \pm 2, \ldots, \pm\frac{N-2}{4}. \tag{4.9}$$

Substituting (4.7) into (4.2) we obtain a quadratic equation for the energy ε:

$$(\varepsilon-\alpha_1)(\varepsilon-\alpha_2)-\beta_1^2-\beta_2^2-2\beta_1\beta_2\cos\frac{4p\pi}{N}=0. \tag{4.10}$$

For each value of p we find two values of the energy

$$\varepsilon_p^{(1)}, \varepsilon_p^{(2)}=\frac{\alpha_1+\alpha_2}{2}\mp\sqrt{\frac{(\alpha_1-\alpha_2)^2}{4}+\beta_1^2+\beta_2^2+2\beta_1\beta_2\cos\frac{4p\pi}{N}}. \tag{4.11}$$

This equation indicates that the energy levels fall into two groups: $\varepsilon_p^{(1)}$ (upper sign before the root) and $\varepsilon_p^{(2)}$ (lower sign before the root). As the energy does not depend on the sign of p, all levels, except $p = 0$, are doubly degenerate, but when $N/2$ is even, the level $p = N/4$ is also non-degenerate. All levels of the first group, $\varepsilon_p^{(1)}$, lie below the mean Coulomb integral $1/2(\alpha_1+\alpha_2)$, and the $\varepsilon_p^{(2)}$ lie, symmetrically, above this mean value. In the ground state of the molecule, the levels of the first group are filled first, followed (if $M > N$) by the levels of the second group. The resonance integrals β_1 and β_2 are usually negative (β_1, $\beta_2 < 0$) so that their product is positive. From Eq. (4.11) it then follows that the energy levels of the first group increase monotonically (and the levels of the second group decrease monotonically) with an increase in the absolute value of p.

On the basis of (4.11), it is easy to derive an equation for the electronic excitation energy of a neutral molecule ($M = N$) (see Section 3). It is important to note that, in this case, $\Delta\varepsilon$ does not tend to zero as $N \to \infty$, but tends to a finite value

$$\lim_{N\to\infty} \Delta\varepsilon(N) = \sqrt{(\alpha_1 - \alpha_2)^2 + 4(\beta_1 - \beta_2)^2}. \tag{4.12}$$

Note also that if

$$\alpha_1 = \alpha_2 \equiv \alpha; \quad \beta_1 = \beta_2 \equiv \beta \tag{4.13}$$

then, (4.11) leads to the same MO energies for a cyclic molecule as the corresponding formula (3.8) for even N. In fact, from (4.11) and (4.13) we can deduce that

$$\varepsilon_p^{(1)},\ \varepsilon_p^{(2)} = \alpha \pm 2\beta \cos \frac{2p\pi}{N}. \tag{4.14}$$

Considering the possible p-values, [Eqs. (4.8) and (4.9)], it is easy to see that the levels of group $\varepsilon_p^{(1)}$ give the lower half, and the levels of group $\varepsilon_p^{(2)}$ the upper half of the whole group of N levels described by (3.8).

Let us now consider the analogous problem for a chain molecule containing an odd number of $N - 1$ atoms. The atoms are assumed to be numbered, with indices running from $k = 0$ to $N - 2$. It is desired to solve Eq. (4.1) with the boundary conditions

$$c_{-1} = 0; \quad c_{N-1} = 0. \tag{4.15}$$

On rewriting in terms of d_l, the problem reduces to the solution of (4.2) with the boundary conditions

$$d_{-1} = 0; \quad d_{\frac{N}{2}-1} = 0. \tag{4.16}$$

The solution is found in an elementary manner (see the case of a chain of identical atoms with full equalization of bonds, Section 3):

$$d_l = A_p \sin \frac{2(l+1)p\pi}{N}. \tag{4.17}$$

Linearly independent solutions are obtained at the following values of p:

$$p = 1, 2, \ldots, \frac{N}{2} - 1. \tag{4.18}$$

Substitution of (4.18) into (4.2) gives the dependence of the energy on p:

$$\varepsilon_p^{(1)}, \varepsilon_p^{(2)} = \frac{\alpha_1 + \alpha_2}{2} \mp \sqrt{\frac{(\alpha_1 - \alpha_2)^2}{4} + \beta_1^2 + \beta_2^2 + 2\beta_1\beta_2 \cos \frac{4p\pi}{N}}. \tag{4.19}$$

The upper sign in front of the root corresponds to the first group of levels $\varepsilon_p^{(1)}$, and the lower sign to $\varepsilon_p^{(2)}$. Taking into account the possible values of p (4.18), this equation determines $N - 2$ energy levels. However, the total number of energy levels should be equal to $N - 1$, which is the total number of atoms. To find the level not given by (4.18) at the above p, we turn to operations related to transition from Eqs. (4.1) to (4.2). It is easy to see that no such transition is possible if the energy has the value

$$\varepsilon = \alpha_2, \tag{4.20}$$

since in this instance the coefficients c_{2l} with even subscripts cannot be expressed by means of the first equation of (4.1) in terms of the coefficients with odd subscripts c_{2l+1}. At the same time, the energy given by (4.20) allows us to find orbital coefficients, which are not all zero and which satisfy (4.1) and (4.15). These coefficients are zero with odd subscripts:

$$c_{2l+1}=0 \quad \text{for } \varepsilon=\alpha_2, \tag{4.21}$$

while their values at atoms with even subscripts are, to within the normalizing coefficients,

$$c_{2l}=(-\beta_2/\beta_1)^l. \tag{4.22}$$

This means that in addition to the N - 2 levels found above, we have a level $\varepsilon=\alpha_2$.

In conclusion, note that the problems considered in this Section can also be easily solved if we take the non-orthogonality integrals of the wave functions of neighboring atoms into account. Let the bonds characterized by resonance integrals β_1 and β_2 be also characterized by non-zero non-orthogonality integrals S_1 and S_2. Setting up Eqs. (1.7) for this case and repeating all the transformations carried out above (when we assumed $S_1=S_2=0$), we can show that in a ring containing an even number N of atoms or in a chain containing an odd number N - 1 of atoms, the energy levels are given by

$$\varepsilon_p^{(1,2)}=(B\mp\sqrt{B^2+4AC})/2A, \tag{4.23}$$

where

$$A=1+S_1^2+S_2^2+2S_1S_2\cos\frac{4p\pi}{N}, \tag{4.24}$$

$$B=\alpha_1+\alpha_2+2\beta_1S_1+2\beta_2S_2+(\beta_1S_2+\beta_2S_1)\cos\frac{4p\pi}{N}, \tag{4.25}$$

$$C=\alpha_1\alpha_2+\beta_1^2+\beta_2^2+2\beta_1\beta_2\cos\frac{4p\pi}{N}. \tag{4.26}$$

The possible values of p for the ring and for the chain are again given by (4.8), (4.9) and (4.18), and the chain also has an additional "special" level $\varepsilon=\alpha_2$.

5. CYCLIC MOLECULE IN A MAGNETIC FIELD

Consider a cyclic molecule similar to the one discussed in Section 4, which contains an even number, N, of atoms, and which is situated in a homogeneous magnetic field perpendicular to the plane of the ring. Without loss of generality, we can regard the ring as a regular N-sided polygon. (This is a consequence of London's approximation [6] used below; in this approximation the energy of the electrons of the ring in a magnetic field depends only on the area of the ring and not on the details of its geometry.)

In accordance with London's theory [6], the orbital coefficients of the MO's in a magnetic field are determined by equations analogous to (1.14), in which all resonance integrals β_{kl} acquire phase factors of the form

$$\omega_{kl} = e^{\frac{ieH\sigma_{kl}}{hc}}, \qquad (5.1)$$

where H is the magnetic field intensity, and σ_{kl} is the area of a triangle whose three vertices are the (coordinate) origin and the locations of the kth and lth atoms, and where e, h, and c are fundamental physical constants. Allowing for (5.1), the formulae for the orbital coefficients have the form [see (4.1)]:

$$\begin{aligned} \beta_1\omega^{-1}c_{2l-1} + (\alpha_2 - \varepsilon)\,c_{2l} + \beta_2\omega c_{2l+1} &= 0, \\ \beta_2\omega^{-1}c_{2l} + (\alpha_1 - \varepsilon)\,c_{2l+1} + \beta_1\omega c_{2l+2} &= 0. \end{aligned} \qquad (5.2)$$

The coordinate origin is selected in the center of a regular N-sided polygon. Let us introduce the notation

$$\omega = e^{\frac{ieH\sigma}{Nhc}}, \qquad (5.3)$$

where σ is the area of the ring. As in Section 4, we transform from Eqs. (5.2) to equations

$$\beta_1\beta_2\omega^{-2}d_{l-1} + [\beta_1^2 + \beta_2^2 - (\varepsilon - \alpha_1)(\varepsilon - \alpha_2)]\, d_l + \beta_1\beta_2\omega^2 d_{l+1} = 0, \tag{5.4}$$

which are satisfied by quantities

$$d_l \equiv c_{2l+1} \tag{5.5}$$

with the periodicity conditions

$$c_{k+N} = c_k. \tag{5.6}$$

For the quantities d_l, these conditions are written in the form

$$d_{l+N/2} = d_l. \tag{5.7}$$

The solution of (5.3) with conditions (5.6) is again of the form [see (4.7)]:

$$d_{lp} = C_p e^{\frac{4lp\pi i}{N}}. \tag{5.8}$$

Linearly independent solutions are obtained at the following values of p:

$$\text{for even } N/2: p = 0, \pm 1, \ldots, \pm\left(\frac{N}{4} - 1\right), \frac{N}{4}, \tag{5.9}$$

$$\text{for odd } N/2: p = 0, \pm 1, \ldots, \pm\frac{N-2}{4}. \tag{5.10}$$

Substitution of (5.8) into (5.3) allows us to determine the dependence of the energy on p:

$$(\varepsilon - \alpha_1)(\varepsilon - \alpha_2) - \beta_1^2 - \beta_2^2 - 2\beta_1\beta_2 \cos\frac{1}{N}\left(4p\pi + \frac{2eH\sigma}{hc}\right) = 0. \tag{5.11}$$

Solution of this equation again gives two groups of levels:

$$\varepsilon_p^{(1)}, \varepsilon_p^{(2)} = \frac{\alpha_1 + \alpha_2}{2} \mp \mp\sqrt{\frac{(\alpha_1 - \alpha_2)^2}{4} + \beta_1^2 + \beta_2^2 + 2\beta_1\beta_2 \cos\frac{1}{N}\left(4p\pi + \frac{2eH\sigma}{hc}\right)}. \tag{5.12}$$

Together with the values of p given by (5.9) and (5.10), this formula defines all energy levels of a ring molecule in a magnetic field. In contrast to the case $H = 0$ discussed in Section 4 [see (4.11)], the levels which differ only in the sign of p do not coincide, i.e., the magnetic field lifts the degeneracies.

Differentiation of (5.12) with respect to H gives the contribution of the given energy level to the diamagnetic susceptibility of the molecule:

$$\chi_p^{(1,2)} \equiv -\left(\frac{\partial^2 \varepsilon_p^{(1,2)}}{\partial H^2}\right)_{H=0} =$$

$$= \mp \frac{4e^2\sigma^2\beta_1\beta_2\left\{\left[\frac{(\alpha_1-\alpha_2)^2}{4}+\beta_1^2+\beta_2^2\right]\cos\frac{4p\pi}{N}+\beta_1\beta_2\left(1+\cos^2\frac{4p\pi}{N}\right)\right\}}{h^2c^2\left[\frac{(\alpha_1-\alpha_2)^2}{4}+\beta_1^2+\beta_2^2+2\beta_1\beta_2\cos\frac{4p\pi}{N}\right]^{3/2}}. \tag{5.13}$$

The upper sign in front of the fraction corresponds to the group of levels $\varepsilon_p^{(1)}$, and the lower to the group $\varepsilon_p^{(2)}$.

6. MOLECULAR ENERGY LEVELS IN POLYACENES

As the next example of the application of the method of finite differences we shall consider the linear polyacenes. The structure of such a molecule is shown diagramatically in Fig. 2, which indicates the numbering system for the values of integrals α and β.

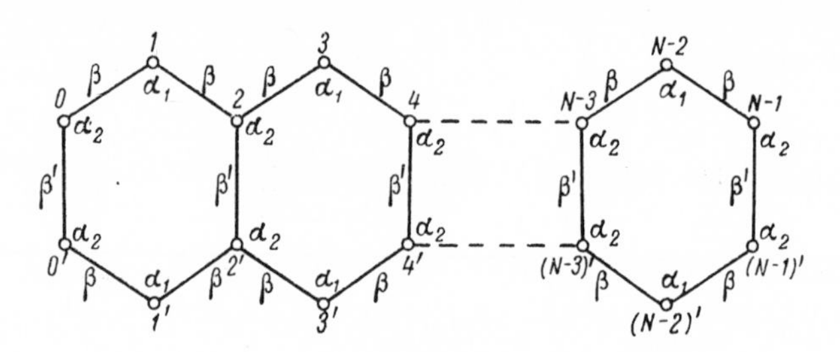

Fig. 2

It can be seen from Fig. 2 that in the chains of bonds 0—1—2—... and 0′—1′—2′—... the resonance intergrals of all bonds are equal, while the Coulomb integrals alternate, taking on a value of α_2 at atoms with even subscripts and a value of α_1 at atoms with odd subscripts. The resonance integrals of the "vertical" bonds $0-0'$, $1-1'$, etc., are assumed to be β'.

From the symmetry of the molecule it follows that the orbital coefficients c_k and $c_{k'}$ should either coincide (symmetrical MO's, denoted by S) or should differ only in their sign (antisymmetrical MO's, denoted by A):

$$\begin{aligned} &\text{case } S\text{: } c_{k'} = c_k; \\ &\text{case } A\text{: } c_{k'} = -c_k. \end{aligned} \tag{6.1}$$

In other words, only the coefficients referring to atoms carrying unprimed subscripts are independent. Equations (1.14) for these coefficients are of the form:

$$\begin{aligned} \beta c_{2l} + (\alpha_1 - \varepsilon)\, c_{2l+1} + \beta c_{2l+2} = 0, \\ \beta c_{2l-1} + (\alpha_2 - \varepsilon \pm \beta')\, c_{2l} + \beta c_{2l+1} = 0. \end{aligned} \tag{6.2}$$

Both here and in further discussion, the upper sign on β' refers to states S and the lower sign to states A. Exclusion of coefficients with even subscripts from (6.2) gives:

$$\beta^2 d_{l-1} + [2\beta^2 - (\alpha_1 - \varepsilon)(\alpha_2 - \varepsilon \pm \beta')]\, d_l + \beta^2 d_{l+1} = 0. \tag{6.3}$$

We now put

$$d_l \equiv c_{2l+1}. \tag{6.4}$$

The orbital coefficients c_k satisfy the boundary conditions

$$c_{-1} = 0; \; c_N = 0, \tag{6.5}$$

which for quantities d_l are written in the form

$$d_{-1} = 0; \; d_{\frac{N-1}{2}} = 0. \tag{6.6}$$

The solution of (6.3) with conditions (6.6) is:

$$d_{lp} = A_p \sin \frac{2(l+1)\,p\pi}{N+1}. \tag{6.7}$$

The following values of p correspond to linearly independent values:

$$p = 1, 2, \ldots, \frac{N-1}{2}. \qquad (6.8)$$

Substitution of (6.7) into (6.3) allows us to determine the energy

$$\varepsilon_p = \frac{\alpha_1 + \alpha_2 \pm \beta'}{2} \mp \frac{1}{2} \sqrt{(\alpha_1 - \alpha_2 \mp \beta')^2 + 16\beta^2 \cos^2 \frac{p\pi}{N+1}}. \qquad (6.9)$$

The upper sign before β' corresponds to states S and the lower sign to states A. Moreover, two choices of the sign in front of the root are possible. Formulae (6.8) and (6.9) determine $2(N-1)$ energy levels. The total number of levels should, however, be equal to the number of atoms, i.e., $2N$. The remaining two levels are found by careful inspection of the transition from (6.2) to (6.3). During this transition we can "lose" levels for which the coefficient preceding c_{2l} in the second equation of (6.2) is zero, i.e., levels for which

$$\varepsilon = \alpha_2 \pm \beta'. \qquad (6.10)$$

We find that non-zero values of c_k satisfying equations (6.2) and boundary conditions (6.5) and with energies shown in the above formula do, in fact, exist. Among these, all coefficients with odd subscripts are equal to zero,

$$c_{2l+1} = 0, \qquad (6.11)$$

while the normalized orbital coefficients with even subscripts are

$$c_{2l} = \frac{(-1)^l}{\sqrt{N-1}}. \qquad (6.12)$$

The two signs in (6.10) correspond to the two types of symmetry (S and A).

The progressively more complicated examples considered in this chapter are a good illustration of the possibilities of the method of finite differences. In every case, we were able to solve the problem of electronic energy levels in the molecule using the simple MO method. The solutions were easy to obtain and were complete. Moreover, it would not be difficult to generalize the formulas of this section to include the case in which the non-orthogonality integrals are included (exactly as in Section 4).

7. AN INFINITE CHAIN OF ATOMS. ENERGY ZONES AND LOCAL STATES

The method of finite differences can also provide information about energy zones in infinite structures. First of all, we turn to the case of an infinite one-dimensional chain of atoms, assuming that the chain has the structure shown in Fig. 1.

The orbital coefficients are given by Eqs. (4.1)-(4.3). In addition, it follows that from the translational symmetry of the chain with respect to a displacement by the sum of the lengths of two adjacent bonds, that

$$c_{k+2} = c_k e^{i\varkappa} \quad \text{or} \quad d_{l+1} = d_l e^{i\varkappa}, \tag{7.1}$$

where $\varkappa$ is real. This condition, together with Eq. (4.3), is satisfied by coefficients d_l of the form

$$d_l = e^{il\varkappa}, \tag{7.2}$$

and the dependence of the energy ε on $\varkappa$ is again determined by substituting (7.2) into (4.2).

$$\varepsilon^{(1)}(\varkappa),\ \varepsilon^{(2)}(\varkappa) = \frac{\alpha_1 + \alpha_2}{2} \mp \sqrt{\frac{(\alpha_1 - \alpha_2)^2}{4} + \beta_1^2 + \beta_2^2 + 2\beta_1\beta_2 \cos\varkappa}. \tag{7.3}$$

Linearly independent solutions correspond to values of $\varkappa$ in the interval

$$-\pi < \varkappa \leqslant \pi. \tag{7.4}$$

The energy values are grouped into two zones, $\varepsilon^{(1)}$ and $\varepsilon^{(2)}$, which correspond to the upper and lower signs in front of the root in (7.3). The energy difference between the lowest level of the upper zone $\varepsilon^{(2)}$ and the highest level of the lower zone $\varepsilon^{(1)}$ is

$$\Delta\varepsilon = \sqrt{(\alpha_1 - \alpha_2)^2 + 4(\beta_1 - \beta_2)^2}. \tag{7.5}$$

The zones touch ($\Delta\varepsilon = 0$) only when

$$\alpha_1 = \alpha_2 \equiv \alpha; \quad \beta_1 = \beta_2 \equiv \beta. \tag{7.6}$$

In this case we have a simple chain of identical atoms joined by fully equalized bonds. Such a chain exhibits translational symmetry with respect to displacement by a length of one bond. For the orbital coefficients this reduces to the following condition

$$c_{k+1} = c_k e^{i\lambda}, \text{ where } \lambda \text{ is real} \tag{7.7}$$

Condition (7.7) and Eqs. (3.1) are satisfied by the expression

$$c_k = e^{ik\lambda}, \tag{7.8}$$

where λ can vary between $-\pi$ and π.

$$-\pi < \lambda \leqslant \pi. \tag{7.9}$$

For the energy, we obtain

$$\varepsilon(\lambda) = \alpha + 2\beta\cos\lambda. \tag{7.10}$$

Equation (7.3) derived above expresses the splitting of the energy zone (7.10) into $\varepsilon^{(1)}$ and $\varepsilon^{(2)}$, occurring as a result of the alternation of the integrals α and β.

We shall now show that the method of finite differences can also be used to investigate the appearance of local electronic states related to local disturbances of the periodicity. We shall consider the simplest cases of disturbance of periodicity: a local change in the value of integral α and a local change in the value of integral β.

Consider an infinite chain of atoms with indices $k = 0, \pm 1, \pm 2, \ldots$, and let the Coulomb integrals of all atoms with index

$k \neq 0$ be α, and the Coulomb integral of the atom with index $k = 0$ be α_0. The resonance integrals of all bonds are assumed to be equal to (β). The problem is to find local states of the electrons (if these exist), i.e., states whose wave functions tend to zero with increasing distance from the atom whose index $k = 0$. This condition reduces to

$$c_k \to 0 \quad \text{at} \quad |k| \to \infty. \tag{7.11}$$

At $|k| \geqslant 1$, the orbital coefficients satisfy the equation

$$\beta c_{k-1} + (\alpha - \varepsilon) c_k + \beta c_{k+1} = 0. \tag{7.12}$$

Moreover, the equation

$$\beta c_{-1} + (\alpha_0 - \varepsilon) c_0 + \beta c_1 = 0. \tag{7.13}$$

should be satisfied.

In the example under consideration, there is symmetry with respect to inversion through the point $k = 0$. For states antisymmetric (A) with rspect to this operation,

$$c_{-k} = -c_k. \tag{7.14}$$

Hence, only the coefficients c_k with positive subscripts are independent, and $c_0 = 0$ for states of type A. Equation (7.13) is satisfied, and the presence of a change in the value of the Coulomb integral for the atom $k = 0$ does not affect Eq. (7.12) which is for $|k| \gg 1$. The solutions in case A are therefore given by the formula

$$c_k(\lambda) = \sin k\lambda, \tag{7.15}$$

and there are no local states. For the symmetric states S we have

$$c_{-k} = c_k. \tag{7.16}$$

The c_k with subscripts $0, 1, 2, \ldots$, which must satisfy (7.12) and (7.13), are independent. Taking (7.16) into account, we can write (7.13) in the form

$$(\alpha_0 - \varepsilon) c_0 + 2\beta c_1 = 0. \tag{7.17}$$

We must now solve Eqs. (7.12) with condition (7.17). We seek localized states, and, therefore, in the general solution of (7.12) at $k \geqslant 1$

$$c_k(\lambda) = C_+ e^{k\lambda} + C_- e^{-k\lambda}, \ \mathrm{Re}\,\lambda > 0 \tag{7.18}$$

C_+ must be made equal to 0. This gives

$$c_k(\lambda) = C_- e^{-k\lambda}. \tag{7.19}$$

Substitution of (7.19) into (7.12) gives the energy of a local state in terms of the parameter λ:

$$\varepsilon_0 = \alpha + 2\beta \cosh \lambda \tag{7.20}$$

This equation shows that a local level (if it exists) lies below all the zone energy levels determined by (7.10) (remember that $\beta < 0$). It follows from Eqs. (7.17), (7.19) and (7.20), that

$$\sinh \lambda = \frac{\alpha_0 - \alpha}{2\beta}. \tag{7.21}$$

Equation (7.21) has a solution $\lambda > 0$ only if $\alpha_0 < \alpha$. The value of sinh λ is then uniquely determined, and consequently so is the energy of the single local level, which, with the aid of (7.20) and (7.21), is written as

$$\varepsilon_0 = \alpha - \sqrt{4\beta^2 + (\alpha - \alpha_0)^2}, \quad \alpha_0 < \alpha. \tag{7.22}$$

(We used the identity $\cosh^2\lambda - \sinh^2\lambda = 1$.)

We now consider another simple type of local periodicity disturbance. Let the Coulomb integrals of all atoms be α, and the resonance integrals of all bonds except the one between atoms 0 and -1 be β. Let the resonance integral of the bond between atoms 0 and 1 be denoted by β_0. From the symmetry of the chain.

$$c_{-k} = \pm c_{k-1}. \tag{7.23}$$

Both here and in further discussion the upper sign refers to states S and the lower to states A. We must now solve (7.12) for $k \geqslant 1$

with the condition that

$$(\alpha \pm \beta_0 - \varepsilon) c_0 + \beta c_1 = 0. \tag{7.24}$$

For the local state, the solution is again of form (7.19), which gives relation (7.20) between the energy of the local state and the parameter λ. With the aid of (7.19) and (7.20), Eq. (7.24) reduces to the condition

$$e^{\lambda} = \pm (\beta_0/\beta). \tag{7.25}$$

Now, $\beta_0/\beta = |\beta_0|/|\beta|$, because β, $\beta_0 < 0$. Moreover, $\lambda > 0$ as above. This is possible only for states S [upper sign in (7.25)] when $|\beta_0| > |\beta|$. Equations (7.20) and (7.25) thus yield the following expression for the energy of the local state:

$$\varepsilon = \alpha + \beta\left(\frac{\beta_0}{\beta} + \frac{\beta}{\beta_0}\right). \tag{7.26}$$

8. APPLICATION OF THE METHOD OF FINITE DIFFERENCES TO TWO-DIMENSIONAL REGULAR STRUCTURES

The method of finite differences can also be applied to two- (or three-) dimensional problems, in which the orbital coefficients depend on two (or three) indices. However, simple solutions are only possible when the difference equations and the boundary conditions permit separation of variables. (Note the analogy to partial differential equations.)

Let us consider some two-dimensional examples. Two-dimensional separation of variables means that the orbital coefficients are written in the form

$$c_{kl} = f(k)\, g(l), \tag{8.1}$$

where $f(k)$ and $g(l)$ are functions of integers k and l.

Consider a rectangular network having m atoms in each horizontal row and n atoms in each vertical column. The Coulomb integrals of all the $N = mn$ atoms are taken to be α. The resonance integrals are taken to be β_1 for horizontal bonds, and β_2 for vertical bonds. The orbital coefficients are c_{kl} where $k = 1, 2, \ldots, m$; $l = 1, 2, \ldots, n$.

The equations for these coefficients have the form

$$(\alpha - \varepsilon) c_{kl} + \beta_1 (c_{k-1,\, l} + c_{k+1,\, l}) + \beta_2 (c_{k,\, l-1} + c_{k,\, l+1}) = 0. \quad (8.2)$$

The boundary conditions at the edges of the network are

$$c_{0,\, l} = c_{m+1,\, l} = c_{k,\, 0} = c_{k,\, n+1} = 0. \quad (8.3)$$

It is easy to see that the variables separate in the expression for the orbital coefficients. The solution has the form:

$$c_{kl,\, pq} = \frac{2}{\sqrt{(m+1)(n+1)}} \sin \frac{kp\pi}{m+1} \sin \frac{lq\pi}{n+1}. \quad (8.4)$$

and values

$$p = 1, 2, \ldots, m; \; q = 1, 2, \ldots, n. \quad (8.5)$$

correspond to linearly independent solutions. Substitution of (8.4) into (8.2) gives the energy in terms of p and q:

$$\varepsilon_{pq} = \alpha + 2\beta_1 \cos \frac{p\pi}{m+1} + 2\beta_2 \cos \frac{q\pi}{n+1}. \quad (8.6)$$

We shall now consider an infinite network. In this case, indices k and l take on all integral values, and there exists a translational symmetry:

$$c_{k+1,\, l} = e^{i\varkappa} c_{kl}; \; c_{k,\, l+1} = e^{i\lambda} c_{kl}. \quad (8.7)$$

The solution of Eq. (8.2) with conditions (8.7) has the form

$$c_{kl} = e^{i(k\varkappa + l\lambda)}. \quad (8.8)$$

The real numbers $\varkappa$ and λ for linearly independent solutions lie in the range $-\pi < \varkappa, \lambda \leqslant \pi$, and the dependence of the electron energy on these numbers is given by the formula

$$\varepsilon(\varkappa, \lambda) = \alpha + 2\beta_1 \cos\varkappa + 2\beta_2 \cos\lambda. \tag{8.9}$$

As our second example, we chall consider a triangular network of atoms, all of whose "cells" are equilateral triangles. [Note that the problem of MO's in a hexagonal network (such as the laminar graphite lattice) can be reduced to the problem for a triangular network by excluding some of the orbital coefficients from equations of (1.14)]. In such a network, every atom is surrounded by six nearest neighbors (Fig. 3).

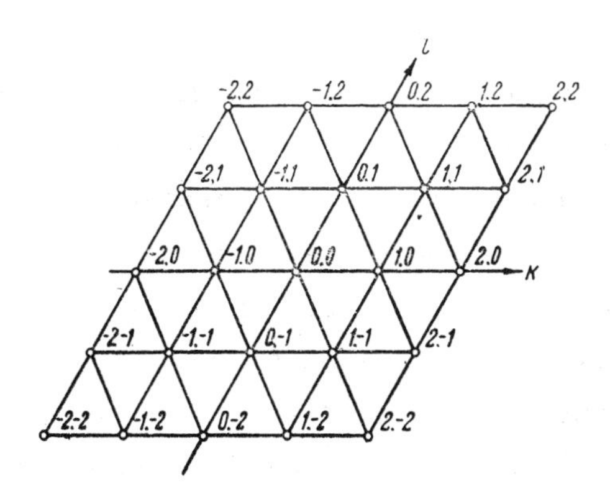

Fig. 3

We shall regard the Coulomb integrals of all atoms to be α and the resonance integrals of all bonds to be β. The equations for orbital coefficients now become:

$$(\alpha - \varepsilon) c_{kl} + \beta (c_{k-1,\, l} + c_{k+1,\, l} + c_{k,\, l-1} + c_{k,\, l+1} + \\ + c_{k+1,\, l-1} + c_{k-1,\, l+1}) = 0. \tag{8.10}$$

The solution for an infinite network satisfying the condition of translational symmetry is easily obtained. It is of the form

$$c_{kl} = e^{i(k\varkappa + l\lambda)}. \tag{8.11}$$

For the energy, we obtain

$$\varepsilon(\varkappa, \lambda) = \alpha + 2\beta [\cos\varkappa + \cos\lambda + \cos(\varkappa - \lambda)]. \tag{8.12}$$

As before, the numbers $\varkappa$ and λ lie in the range

$$-\pi < \varkappa, \ \lambda \leqslant \pi. \tag{8.13}$$

Because of the more complex form of relation (8.12) between the energy and the numbers $\varkappa$ and λ, it is more difficult to satisfy the boundary conditions at the edges of a finite triangular network than it is to satisfy (8.3) for a finite rectangular network. This problem will not be discussed further, and we shall merely note that it has a simple analytical solution only when the boundaries of the network have a very special form.

The very simple examples considered above show that the method of finite differences is useful in calculation of the energy levels for one-dimensional and two-dimensional regular structures. Generalization of this method to three dimensions is not too difficult (assuming separability of the variables). The separation of variables is always possible for infinite periodic structures exhibiting translational symmetry; it is more difficult for finite portions of these structures, since the boundary conditions at the edges may be unfavorable.

CONCLUSIONS

On the basis of Sections 3-8 it is seen that the method of finite difference is an effective tool in the solution of the problems of MO theory for regular structures. In all examples considered we were able to obtain exact solutions to the problem of the energy levels, in simple analytical form. The entire procedure consisted of elementary operations with trigonometric or hyperbolic functions.

Of course, the range of problems in MO theory which may be effectively solved by the method of finite differences is far from exhausted by the examples considered here. A general condition for the applicability of this method is the possibility of dividing the molecule into individual parts each of which have a regular structure. The orbital coefficients are then constructed as above, and conditions ensuring that the basic Eqs. (1.14) are always satisfied as we go from one part to another, must be fulfilled at the boundaries of the individual part of the molecule. For complex structures, the resulting problems can be much more difficult than the examples discussed above, but even then the method of finite differences preserves its advantages over the usual matrix methods, as it ensures an appreciable saving of labor in the calculations.

Finally, it may be noted that one of the factors suggesting the application of the method of finite differences to problems of MO theory was the comparison between the mathematical apparatus of the free-electron model and the method of molecular orbitals [7, 8].

REFERENCES

1. F. Hund. Zs. Phys., 1928, 51, 759.
2. R. Mulliken. Phys. Rev., 1928, 32, 196.
3. E. Hückel. Zs. Phys., 1930, 60, 423.
4. C. A. Coulson and H. C. Longuet-Higgins. Proc. Roy. Soc., 1947, A191, 39; 1947, A192, 16; 1948, A193, 456; 1948, A195, 188.
5. A. O. Gel'fond. Ischislenie Konechnykh Raznostei [Calculus of Finite Differences], Moscow, Fizmatgiz, 1959.
6. F. London. J. Phys. Rad., 1937, 8, 397.
7. T. K. Rebane. Candidate's Thesis, Leningrad State University, 1957.
8. T. K. Rebane. Vestnik Leningrad Gosud. Univ., 1957, No. 22, 70.

INDEX